TRADING WESTERN SOFTWOOD LUMBER

The Basics

DAVE LECKEY

Highland Press
Wilsonville, Oregon

This book is designed to help the reader attain professional competence. Although the material has been reviewed prior to its publication, in dealing with specific transactions, the reader should consult all available authorities. The author makes no expressed or implied warranties regarding the use of the information contained herein. Each reader must depend upon his or her own expertise in making use of these materials.

ISBN 0-9621022-0-2
Library of Congress number: 88-82943

for more copies contact:

Highland Press
P.O. Box 933
Wilsonville, OR 97070

Cover designed by Lynn Guitteau
Illustrations by Jud Guitteau
Text designed by Susan Applegate of Publishers Book Works, Inc.

Acknowledgments

The writing of this book has been somewhat like the writing of an oral history, because so very little has been written and published about the business of trading lumber. For this reason, the author has had to rely on experts in the various aspects of the field to put together much of this information. Many peoples' contributions have been significant. To these colleagues and friends, the author is truly indebted. These people include: Don Bean, Dick Bech, Mick Foy, Leo Gibbons, Jud Guitteau, Dale Gustafson, Bill Hill, Clem Jennings, Terry Lane, Bob McGaughey, Greg Patterson, Evi Nast Selkirk, and Bill Stewart.

A special thanks is due to Cherie Buell, whose proofreading, suggestions, and encouragement have been invaluable.

The final and most important acknowledgment is to my wife, Margaret Beth Neal, without whose help, from inception to publication, this book would not have been possible.

Contents

List of Tables

Introduction

The purpose of this text is to introduce new traders to the business of buying and selling wholesale quantities of western softwood lumber. The material is presented, by chapters, as follows.

Chapter One is about lumber and the lumber market. It starts by introducing what western softwoods are, then discusses the lumber market in North America, and gives a step-by-step description of the sawmill processes that convert a raw log into finished lumber.

Chapter Two introduces the concept of trading. Lumber trading is defined, and the process of trading is described.

Chapter Three talks about the transportation system, since rail, truck, and van transportation are integral parts of lumber trading.

Chapter Four contains information on the mechanics of buying and selling lumber. It introduces the twenty key elements of every western softwood lumber order. Each element is described and its significance is discussed.

Chapter Five is a discussion of some useful techniques that can be used to create orders.

Chapter Six is a discussion of lumber claims; how they can occur, and how to deal with them once they have occurred.

Chapter Seven talks about the lumber futures market and how it can be used as a tool in everyday trading.

The Appendix is a step-by-step look at how to calculate prices on lumber tallies.

Following the appendix is a glossary of terms, page referenced to the text, to cover the unique and specific terminology of the western softwood lumber business.

Finally, a list of study questions (and answers) is provided that refer back to information in the text.

Although the material in this text is aimed primarily at individuals who are interested in becoming lumber traders, it is hoped that others new to any of the various aspects of lumber merchandising--sawmill sales people, specialists in lumber transportation, buyers for distribution or retail lumber yards, or clerical support people--will find this book of use. Additionally, although western softwoods are the focus of this book, much of what will be discussed is applicable to softwoods produced elsewhere in North America.

CHAPTER
ONE

Lumber and the Lumber Market

Before talking meaningfully about lumber trading, one must have some background information about what lumber is, how it is manufactured, and how the lumber marketplace is structured.

WESTERN SOFTWOOD LUMBER--SOME DEFINITIONS

Softwood describes the wood from those trees commonly referred to as conifers or evergreens. Botanically known as gymnosperms, conifers (evergreens) are trees with exposed seeds that are usually protected by cones. Species such as fir, cedar, larch, hemlock, pine, and spruce are all softwoods.

Lumber is wood that has been manufactured from a log and has been sawn on all four sides.

Softwood lumber is favored in so many building and manufacturing applications because: it is relatively plentiful (it is a renewable resource over a time frame of 50 to 150 years); it produces long, straight lumber that can be easily worked with hand tools; it can be readily nailed or glued; it is light in weight relative to its strength; it is a good electrical and thermal insulator; and whether finished or rough, it has appealing visual characteristics.

Western, for the purposes of this text, refers to the geographic region of the United States and Canada running from the Rocky Mountains to the Pacific Ocean. This western region is one of three primary softwood lumber-producing regions of North America (i.e., the United States and Canada), and it is, by far, the largest. The two other principle softwood-producing regions of North America are eastern Canada and the southeastern United States.

Softwood products manufactured in each of the three producing regions are similar and compete with one another in many markets. The primary market for eastern Canadian softwood is eastern Canada and the northeastern United States--from New England to Virginia to Wisconsin. The primary market for southeastern softwoods (southeastern softwoods are predominantly pines, and are referred to generically as "southern yellow pine") is the southeastern quadrant of the United States, in an arc roughly from Pennsylvania to Illinois to Texas. The primary market for western softwoods is all of the United States and Canada.

Board Foot is the primary unit of measure used in the lumber industry. A board foot is a piece of lumber equivalent in size to a one-foot long piece of 1x12. (Read 1x12 as "one by twelve." A 1x12 is a piece of wood 1" thick and 12" wide.) The formula for calculating board footage is:

$$\frac{\text{Thickness (inches) x Width (inches) x Length (feet)}}{12}$$

Using this formula, then, a one foot long 1x12 becomes:

$$\frac{1 \times 12 \times 1}{12} = \frac{12}{12} = 1 \text{ board foot}$$

Work out for yourself that each of the following pieces of lumber contains 1 board foot:

1x6 two feet long
2x6 one foot long
2x8 nine inches long
4x12 three inches long

Thousand Board Feet. Because a board foot is small, and the quantities traders deal in are large, *thousand board feet* is the common unit of measurement in the wholesale lumber industry. It is usually written MBF. A thousand board feet of lumber is simply one board foot times 1,000. Lumber prices and volumes are both discussed in this measurement unit. For example, a trader might discuss with a customer: "28 MBF of 2x4 @ $250/MBF." This reads as "28 thousand board feet of 2 by 4 at a price of $250 per thousand board feet." ("Per thousand board feet" is frequently shortened to read as "per thousand.") 28 MBF is roughly a truckload quantity of lumber , and represents 28 x $250, or $7,000.

Random Length: abbreviated to R/L, this term refers to the lengths of lumber in a particular order when more than two or three lengths are being discussed. It is quite normal to have many different lengths of lumber in an order. In some cases the various lengths are specified, and in some cases they are not (this will be fully discussed later). In the jargon of the industry, people talk about a "2x4 random," or a "1x8 random." "Random" means "random length," and "random length" means an assortment of normal lengths.

THE LUMBER MARKETPLACE

The lumber market of North America is an excellent example of a market that is free and uncontrolled. It is a market in which there are many lumber sellers and many lumber buyers. The amount of lumber produced, the sizes produced, and the sale price, are all determined by the market forces of supply and demand. Each lumber sale is unique and based upon a price negotiated between the buyer and the seller. No prices are fixed, no production is fixed. Much of the lumber produced daily is put on the market without knowing where it will go, and without knowing the sale price that will get it there.

Very little of the western softwood lumber used in North America is purchased directly from the sawmill by the end user. Most lumber works its way through a distribution chain that includes some or all of the following: sawmill sales offices, national lumber trading (office wholesale) companies, regional reload centers, large metropolitan distribution yards, and local lumberyards.

Each link in the lumber distribution chain performs some specific functions, and so each will be briefly described.

Sawmill Sales Office. Traditionally, a sawmill sales office is one or two people selling the production of one, two, or three jointly-owned sawmills. These people are located at the sawmill site, know intimately what the sawmill is producing and can produce, and can make decisions about future production.

These sawmill sales people sell large volumes of lumber to a relatively small number of customers, each with good credit. The customers take care of the complex freight issues, and pay promptly for the lumber after it has been shipped and invoiced. In this way, a sawmill sales office is small, yet capable of moving large volumes of the sawmills' production. There is little need for credit experts or traffic specialists, so the sawmills' sales overhead remains low.

This traditional role of sawmill sales is changing somewhat. As independent sawmills are purchased by larger lumber companies, the newer sawmill sales offices are now selling the production of many sawmills. To increase profits, these larger sawmill sales offices are offering their lumber to more and more buyers. These larger sales offices must have their own traffic experts and expanded credit departments. These offices are often not at a sawmill site, and the salespeople are not as closely connected with production. For these reasons, the operating procedures of a sawmill sales office depends upon the sawmill companies size.

National Lumber Trading (Office Wholesale) Companies. These are companies that market lumber, but are neither the manufacturers of the lumber, nor lumber stocking yards. National lumber trading (office wholesale) companies buy lumber in large quantities from producers (sawmills), then market (sell) it to reload centers, distribution centers, larger independent lumber yards, and to manufacturers who buy lumber in volume. Most of this marketing is done via long distance telephone. These trading and wholesale companies generally employ extensive credit and traffic departments as necessary support for the trading staff. The physical location of these companies is only incidental, because the lumber they trade seldom passes through their location.

Regional Reload Centers. A regional reload center is a lumber handling yard set up to unload lumber from one type of transportation (usually railroad cars), and reload it onto another (usually trucks), for shipment to its final destination. Because this handling involves additional costs, these reload yards are located where rail freight rates are good.

For example, the rail freight rate from western Canada to Niagara Falls, Ontario, Canada, for dry, spruce dimension lumber, is approximately $80 US per MBF. By contrast, shipping the same lumber to Niagara Falls, NY, just 5 miles away but across the US-Canadian border, costs approximately $115 US per MBF. It does not cost the railroad $35 US per MBF to haul the lumber those additional 5 miles; it is, rather, the way that the rail freight tariffs are written that causes the difference. Similar freight anomalies exist at other locations. Reload centers take advantage of such freight advantages.

Reload centers buy lumber in the west, have it shipped by rail to their facility, unload it from the railroad cars, hold it as inventory, then sell it as truckloads delivered (by truck) to final destinations. Depending upon the particular site, reload centers may be able to economically ship truckloads of lumber up to 300 or 400 miles from their location and compete head-to-head with rail direct shipments.

These regional reload centers buy their lumber directly from sawmills and from national lumber trading (office wholesale) companies.

Large Metropolitan Distribution Yards. Large metropolitan distribution yards are like reload centers on a smaller scale. These yards will buy railroad carloads and truckloads of lumber for their own inventory. In addition to lumber, these yards will carry many other building products. From these large inventories, the metropolitan distribution yard will sell truckloads of mixed items to the independent retail lumber yards, contractors, and manufacturers in the area.

Metropolitan distribution yards buy their lumber from sawmills, national lumber trading (office wholesale) companies, and reload centers.

Local Lumber Yards. Local lumber yards are retail yards that carry lumber and other assorted building products. These yards serve independent local contractors, remodelers, and do-it-yourselfers. These yards typically buy lumber in less than straight truckload quantities.

Local retail yards buy their lumber from the large distribution yards and the reload centers.

THE MANUFACTURING PROCESS

Before softwood can be utilized for the various products demanded from it, it must be converted from tree form into lumber. Lumber, remember, is wood that has been manufactured from a log and sawn on all four sides. This manufacturing operation takes place at a sawmill.[1] Some general knowledge of what happens in a sawmill will be useful to better understand the finished lumber products that are bought and sold. Also, by knowing the sequence of production, a trader will more fully understand what can and cannot be done with lumber and its packaging. What follows is a simplified step-by-step description of how a log becomes lumber as it moves through a mill. There are many different levels of machining and computer technology that a particular mill may or may not employ. The flow as described here is simplified for clarity.

As will be discussed more fully later in the text, some lumber is dried and some is not, some is surfaced (dressed) and some is not, some is gradestamped and some is not, etc. A particular finished piece of lumber may or may not go through all of the steps that a sawmill can perform. The basic flow through the mill, however, is the same for all lumber.

The Steps: From Logs to Lumber

Step One. In the forest, trees are felled and bucked (cut to length) for transporting to the sawmill. The length of the logs may be anywhere from 35' to 125', depending upon the hauling equipment and whether the logs are being

---

[1] What is commonly referred to as the "lumber mill" or the "sawmill" is, technically, both (a) a sawmill, where logs are made into rough lumber (the logs are "cut"), and (b) a planer mill, where rough lumber is surfaced into finished lumber (the lumber is "run").

taken over public roads (where there are load length restrictions), or private roads.

The downed trees are delimbed and sorted by species or species grouping. (Some logs from different species are manufactured and sold together due to similar characteristics and strength.) The logs are then loaded and transported to the sawmill.

Step Two. The logs are delivered to the sawmill and stacked in the log deck (a holding yard) by species or species grouping. Logs may stay in this log deck from two days to six months, depending upon the sawmill's log supply, the weather, and the terrain of the logging area. Some sawmills can harvest logs almost year round, others are restricted to the frozen months of winter and the dry months of summer.

Step Three. Logs by species or species grouping are next moved to a staging area.

Step Four. One by one, the logs go through a *debarker* for bark removal.

Step Five. If the sawmill is equipped with electronic or laser scanners, each log is scanned for defects and for its maximum cutting yield. Depending upon market conditions and the types of presold orders, the sawmill has a priority listing of sizes and grades of lumber that it is trying to produce.

Step Six. Each log is still uncut. Depending on defects and taper, the next step is to buck the log to slightly over the longest length (normally, 16' or 20') that will be cut from it.

Step Seven. Each log then goes through the *headrig* (the primary saw), where the rounded sides are cut off and the log is broken down (sawn) into square and rectangular *cants* (large flat-sided timbers). The head rig may be computer controlled, working from the information gathered by the electronic scanner, or it may be manually controlled. If manually controlled, the person (called the sawyer) operating the head rig decides how best to cut the log to get the maximum dollar return for the sawmill.

Step Eight. The cants from the head rig move on conveyor belts to secondary saws called *gang saws* and *edgers*. A gang saw looks like a giant bread slicing machine. As a cant moves through the gang saw, multiple blades cut it into numerous smaller pieces. For example, a 16"x12" cant might be cut into eight pieces of 2x12.

From the gang saw, the lumber goes to the edger, where irregular edges and defects are trimmed off. The edger leaves each piece of wood rectangular and in its final rough thickness and width size.

The wood at this stage is called *rough, green lumber.*

Step Nine. The rough, green lumber is separated into piles of lumber all the same thickness and width. If the lumber is to be dried, it goes to Step Ten; if not, it goes to Step Twelve.

Step Ten. To prepare the lumber for drying, the piles of the same thickness and width, rough, green lumber are stacked into units. Between each tier of a unit, and perpendicular to the lumber, half-inch wooden spacers are placed 2' to 2-1/2' apart. These spacers create a gap between each of the lumber tiers. This spacing process is called *stickering*, and allows for increased air circulation to each piece of lumber within the unit. These stickered units then go either to the kiln for drying or outside for air drying.

Step Eleven. The length of time taken to dry the rough, green lumber depends upon many things: the species, the width and thickness of the lumber, the time of year, and the desired quality and dryness of the finished product. On average, lumber usually stays in a drying kiln from twenty-four to seventy-two hours. Lumber that is being air dried, can take as little as three to four weeks, or as long as four to six months.

Once lumber has reached the desired moisture content, it is called, *rough, dry lumber.*

Step Twelve. The rough, dry lumber comes out of the kiln, or off the air-dry yard, and is taken to the planer mill. The *planer* is a machine that gives all four sides of the lumber a smooth finish and makes the lumber absolutely uniform in width and thickness. It is because only one size of lumber can go through the planer at a time that the lumber has already been sorted into the same thickness and width units

The lumber at this point is called *surfaced, dry lumber* if dried, or *surfaced, green lumber* if not dried.

Step Thirteen. The surfaced lumber is immediately graded. The person doing the grading looks at each piece of lumber separately and determines its grade. Depending upon the sawmill's setup, the grader may either designate the grade in crayon near the end of the board, or may physically stamp the lumber with an inked gradestamp. As part of determining the grade, the grader also determines how much to trim off one or both ends to maximize each particular piece's value. This trim designation is also marked in crayon near the end of the board.

Step Fourteen. The lumber passes by conveyor belt to the automatic moisture meter (optional) where its moisture content is checked.[2] The lumber then goes past the mechanical gradestamper (also optional), to the end-trim saws where the graders' end-trim instructions are carried out and the lumber is trimmed to its final length.

The manufacturing of the lumber is finished at this point.

Step Fifteen. The now-finished lumber, which has already been sorted for thickness and width, is further sorted by length (or multiple lengths if more than one length is packaged together), and by grade. Each different length and grade is placed in a separate sorting bin.

Step Sixteen. Taken from the sorting bins, the lumber--all being the same grade, thickness, width, and length--is steel-banded together into finished units. (This is called *unitizing*). Some sawmills paint or wax the ends of their units, and would do that now.

Step Seventeen. The units of lumber are paper wrapped (optional) and moved by forklift to a holding yard or a dry storage shed to await shipment.

[2] At some mills, the moisture meter is before the planer, thus enabling non-dry rough lumber to be re-stickered and re-dried before being planed.

CHAPTER
TWO

Trading

Western softwood lumber trading is a practical business response to the economic realities of supply and demand in the western softwood lumber industry. Because of the free market nature of the lumber industry, no one operating in the market is guaranteed an income or a profit. Income and profit are generated only when a product is sold profitably, or a service is rendered that someone else is willing to pay for.

How lumber is bought and sold, and the services that a good lumber trader can perform for suppliers and customers, is the topic of this chapter.

WHAT IS TRADING?

Trading, as the term is used in the lumber business, means the buying, selling, and transporting of wholesale quantities of lumber. The term trading is frequently used interchangeably with *wholesaling*, although the two are not exactly the same. Traders do not usually buy for, or sell from, their own existing physical inventories. Wholesalers do. Instead, traders typically buy and sell in large-volume lots, owning what they trade only a short time, usually selling what they own before it reaches a destination. Traders seldom actually see the material that they are trading.

Brokering is another term frequently confused with trading. Traders and wholesalers actually take ownership of what they buy and sell. Brokers do not. A broker brings a buyer and a seller together, then earns a commission based upon their order.

A lumber trader is an agent of a lumber trading company. A lumber trader's job is to: buy what sawmills want to sell, sell what customers want to buy, arrange for transportation from the sawmill to the destination as economically as possible, and generate a profit while doing this.

Western softwood lumber is traded because it has high price and availability fluctuations. Because lumber prices and availability change from one day to the next, both producers (sawmills) and customers (lumber stocking yards and manufacturers) find that it makes sense to deal with lumber traders who specialize in this volatile market.

As mentioned in the introduction to this chapter, it is only because a necessary service is rendered that a lumber trader can generate income and profit, and therefore exist, in the market. For a sawmill, a good lumber trader provides broader marketing exposure and reduced responsibilities for credit and transportation. For a customer, a good trader provides accurate and timely market information, a supply of well-priced lumber, the extension of credit, and fast, personalized service .

HOW IS TRADING DONE?

Because a trader actually takes ownership of lumber before it is resold, a lumber trade has two distinct halves, that is, two distinct orders. One order is a *buy order* with a sawmill; the other order is a *sell order* with a customer. Both of these orders may be done by the same trader, or they may be done by two different traders working together in the same office.

Depending upon whether market prices are going up or going down, the buy order may come before the sell order, the sell order may come before the buy order, or the buy order and the sell order may happen simultaneously.

If the buy and the sell orders happen at the same time, the trade is called a *back-to-back*. A back-to-back trade is the final outcome of, usually, several phone conversations between the buying trader and the sawmill, and the selling trader and the customer. When, through negotiation, the sawmill, the trader(s), and the customer are all in agreement as to the type of lumber and the details of when and how it is to be shipped, the buying trader puts the buy order together with the mill. This means the buying trader buys the carload/truckload of lumber in question.

What is the buy order? The buy order is a verbal, followed by a written, contract between the sawmill and the trader. In this contract, the sawmill commits to provide, and the trader (and the trader's company) commit to take ownership of, and to pay for, a particular quantity of lumber. The buy order specifies the characteristics of the lumber, the packaging of the lumber, how the lumber is to be shipped, and the price and terms of payment. In a back-to-back order (where the lumber has already been sold), the order also has a destination (the customer's location). The trader gives this destination information to the sawmill so that the lumber can be shipped directly to the customer.

At the same time that the buy order is being put together on a back-to-back, the selling trader is transacting the sell order with the customer.

What is the sell order? Similar to the buy order, the sell order is a verbal, followed by a written, contract between the customer and the trader. The trader agrees to provide, and the customer agrees to accept and to pay for, a particular quantity of lumber. Just as with the buy order, the sell order specifies the type of lumber, its packaging, its shipping, and its pricing information.

A back-to-back trade involves no market risk or market speculation because the buy order and the sell order happen at the same time. Profit on this trade is strictly the difference between the net delivered buying price and the net delivered selling price.

As anyone involved in buying and selling knows, in addition to providing a quality service, the old adage, "buy low, sell high" is the key to generating profit. Because prices on the lumber market change frequently, a trader who feels that market prices are going to go higher will want to "buy low" so that at some time in the future, after the price is up, the opportunity is there to "sell high." In a situation like this, the buy order may be placed days, or weeks, before the sell order. In lumber vernacular, a trader buying first and selling later is said to be *going long* or taking a *long position*.

Conversely, a trader who feels that market prices are going to go lower will sell an order at the current price levels ("sell high"), so that at some time in the future, after market price levels have moved down, the opportunity exists to "buy low." In a situation like this, the sell order is placed some time before the buy order. The trader selling this order is said to be *going short*, or taking a *short position*.

Traders will go long when they think that market prices are going to go higher. They may not know exactly to whom they will sell the lumber, but by owning it the cost is fixed. As the market prices move up, an opportunity is created to sell at the higher price and thereby generate profit.

Traders who thinks that prices are going lower will go short by selling an order days or weeks before actually purchasing it. These orders may be sold at a price equal to or even below current cost, if the trader feels that prices are going to drop quickly. Again, the trader may not know from which sawmill the lumber will be bought, but with the sale price fixed, as market prices move down, the trader has an opportunity to buy the order with profit.

Any time a long or short position is taken, a trader is speculating. He/she is gambling on which direction the market is going to go. Naturally, this, as all speculation, involves some financial risk. For the trader to make a profit, the market must move in the anticipated direction. If the market does not move in that direction, or if it does not move there fast enough, the time constraints built into every long or short order may force the second half of the trade to be completed before the trader's desired profit level is reached. This can result in smaller profit than anticipated, or it can even result in a loss.

The time constraint in a long or unsold order is the following: when a sawmill has lumber for an order ready, the mill wants to ship the lumber so that it can get paid. If the trader does not yet have the order sold, the sawmill must be given a temporary destination called a *diversion point*. A diversion point is a centrally located geographic spot to which unsold lumber can be shipped. Minneapolis, Minnesota; Beloit, Kansas; and North Platte, Nebraska, are a few commonly-used diversion points.

A railroad car shipping to a diversion point is called a *roller*, or a *transit*. Customers wishing prompt shipment are usually interested in these types of railroad carloads, and will sometimes pay a premium to get them.

When a roller is finally sold to a customer, the new destination is given to the railroad and the car is *diverted* or *switched* to its new, final location. If a car should arrive at the diversion point still unsold, it is said to go *on hand*. On hand means that the car will not move farther without additional information from the shipper.

Once a car is on hand, the railroad starts charging a fee for the additional time use of the car. This fee is called *demurrage* (pronounced *di-mur'-ij*). Demurrage usually starts 1 to 3 days after a car arrives at its destination, and the cost accelerates on an upwardly sliding scale. Once a car is on hand, a trader is under great pressure to sell it, to keep the demurrage costs down.

The time constraint on a short order is that the customer is expecting the lumber to ship within the time specified when the order was taken. The lumber must be bought to ship within that time frame or the trader faces the possibility of a claim for late shipment being filed by the customer.

Because of the nature of speculation, **some** of a trader's guesses about the market's direction (up or down) will be wrong. It is necessary for every trader, then, to remember to maximize profitable positions and minimize loss positions at all times. In a rising market, when a trader owns lumber, selling prices **must** rise as the market will allow, thereby increasing profit. A trader must generate this increased profit when possible because in a falling market, when the trader will also own lumber (although wishing not to), selling prices will **have** to be lowered in order to compete with the falling market prices. If a trader does not have profits to offset the losses, the trader will not remain in business over the long run.

CHAPTER THREE

The Transportation Function

As most lumber is sold on a delivered price basis (with the transportation charges included in the sales price), it is essential to have a good working knowledge of the various lumber transportation systems. There are four major systems used for hauling lumber; they are railroads, trucks, piggyback vans, and ocean barges.

This chapter will explore the basic concepts of operation, rate structure, and the terminology for three of these four modes of transportation. Specifically, we will explore the railroad system, the over-the-road truck system, and the piggyback van system. The fourth mode of transportation, ocean-going barges, is highly specialized and outside the realm of day-to-day lumber trading.

RAILROAD TRANSPORTATION

Railroads are like privately-owned toll roads. To bulk shippers, they offer the advantage of relatively cheap and relatively fast transportation between certain locations. Because they are private and are operated to make a profit, railroad companies build and maintain track in those areas where they can expect to haul the most goods. Access, then, is restricted. Origins and destinations are limited to those points located along railroad tracks.

Almost every fully-operational sawmill is located on, or has access to, a railroad *siding*. A siding is a short, dead-end track coming off the main rail line; it is used for the loading and unloading of railroad cars. The railroad company (or *carrier*) that owns and supplies service to this siding is called the *originating carrier* of any lumber shipping by railroad from that sawmill. Occasionally, a sawmill will have sidings served by more than one railroad company. This enables the sawmill to take advantage of different tariff rates to different destinations.

Similarly, customers who receive railroad carload shipments must be located on a railroad siding. The railroad company that owns this siding, or the access to it, is called the customer's *delivering carrier*. A customer not located on, or having access to, a rail siding may still receive carload deliveries. This customer, however, must have the railcars shipped to the railroads' public siding, called a *team track*. The customer taking rail delivery at a team track must take a forklift and truck to the public siding to unload the car.

In moving lumber from sawmills to destinations, it is frequently necessary to ship lumber to points other than those served by a single railroad, or to buy lumber from mills located on railroads not the same as the customer's. In order to do this, the loaded railcar must be switched from one railroad companies tracks to another. This switching of carriers can only take place at interconnecting points, or *interchanges*, served by both railroads. To get from a sawmill on one railroad to a customer on another may require several interchanges, and several different *intermediate* railroads.

A user's fee is charged by each different railroad company for travel on its system. This fee is usually determined by the length of the trip and the total weight of the lumber being hauled.

The freight charge for an entire trip may take the form of a single freight bill, including each railroad's portion of the haul, or the charges may come as two or three separate freight bills from the individual carriers.

Freight bills may be paid to the originating railroad when a car has shipped, or it may be paid to the terminating railroad upon arrival at the destination. If the freight bill is issued by the originating carrier, then the billing is said to be *freight prepaid*. If the freight is to be paid upon arrival at destination, the bill is issued by the delivering carrier and the billing is said to be *freight collect*.

Whether a shipment is going freight prepaid or freight collect, a *bill of lading* is made out at the time of shipment and travels with the railcar to its destination. Included on the bill of lading is a complete listing of all of the railroads that will carry the car, and a description of exactly where interchanges are to take place. This itinerary is called the car's *routing*.

Because there are a number of railroad companies owning track and competing for business, it is possible to have more than one route from a given sawmill to a given destination. Also, because each of these railroad companies set their own tariff rates (with the exception that some rates are regulated by the United States and the Canadian governments), different routes may result in different rates.

Most tariff rates are published by the railroad companies and, for any particular origin to a particular destination, are the same for every shipper. Occasionally, however, railroads will sign special contracts with large-volume shippers (the shipper is whoever is paying the freight bill), giving these shippers special rates cheaper than the published tariff rates. These special rates are called *contract rates*, and are used by railroads as a competitive tool to increase their market share in a particular area.

While some users (sawmills primarily) of the railroad systems own or lease their own rail equipment (railcars), the great majority of all rail equipment is owned or controlled by the railroad companies. There are many different types of railcars. The railroads provide these cars to shippers for the hauling of all the various things (not just lumber) that railroads carry. Because the different types of equipment have different capacities and have competing uses for their services, the railroads establish a minimum amount of weight that must be shipped on each type and size of car. As most rates are a function of the weight being shipped, this minimum-weight provision guarantees the railroad a certain level of revenue with each use of the car.

Because rates, and therefore revenues, are usually based upon the weight being shipped, railroads encourage shippers to increase the weight per car with a rate structure that favors heavy shipments. These rates are called *incentive* rates, and are characterized by a basic rate for the minimum weight, then a reduced rate for any additional weight above that minimum. This reduced rate is called the *excess* rate. These weight-based rates are quoted in terms of a price *per hundredweight*. (Per hundredweight means per 100 pounds and is abbreviated as "cwt.")

Let's look at an example. The rate from Spokane, Washington to Chicago, Illinois might be quoted as "3.45 on 85, 3.20 on the excess." This means that the railroad is going to charge $3.45 per cwt (or $3.45 for every 100 pounds) on the first 85,000 pounds shipped (this is the minimum rate, regardless of the actual weight shipped), and $3.20 per cwt on everything over 85,000 pounds. If we can suppose that a car has just been shipped on this rate, and that the weight of the lumber in this car is 115,000 pounds, then the railroad's bill will look as follows:

85,000 lbs. @ $3.45/cwt	$2,932.50
30,000 lbs. @ $3.20/cwt	$ 960.00
Total due	$3,892.50

Sometimes a rate will be quoted in two parts. The first part will be given as a rate to a particular point, the second part is another rate *beyond* to the destination. These two-part rates are called *combination rates*.

Here is an example of a combination rate. Going from Columbia Falls, Montana to Chattanooga, Tennessee, the rate might be: "3.60 on 85, 3.35 on the excess, to East St. Louis; $1,226 per car beyond."

This means that it will cost $3.60 per cwt on the first 85,000 pounds, and $3.35 per cwt on the excess, to get the car as far as East St. Louis, Illinois. From there, the beyond portion of the rate is quoted as a *per car* charge of $1,226. (A per car charge, although less frequently seen than weight-based rates, occurs regularly and simply means that a flat fee is charged for the haul, regardless of the weight or footage in the car.)

In this case, let's assume that we are shipping a car weighing 105,000 pounds that contains 69,550 board feet of lumber. We can figure the rate as follows:

From Columbia Falls to East St. Louis:	
85,000 lbs. @ $3.60/cwt	$3,060.00
20,000 lbs. @ $3.35/cwt	$ 670.00
sub-total	$3,730.00
From East St. Louis to Chattanooga:	
Per car charge	$1,226.00
Total Charge	$4,956.00

Lumber prices are usually quoted delivered to the destination, and in terms of a price per MBF. When quoting lumber using this rate, one would take the total freight ($4,956), divide by the footage being shipped (69.55), and arrive at a freight price/MBF. In this case, the freight/MBF would amount to $4,956/69.55 = $71.25/MBF (always round to the nearest $.25/MBF) This freight price is added to the cost of the lumber (per MBF) and the profit (per MBF) to determine the total delivered selling price (per MBF).

Weights

Because weights are very important in determining most railroad freight rates, and because not all parties shipping lumber treat weights the same way, weights deserve some special attention here.

Some sawmills shipping lumber by rail attempt to make money on the freight to the destination as well as on the lumber that they are selling. This extra revenue is generated by the way that these mills deal with the weights of the lumber. It is important to know, at the time of buying, how the weights are to be handled.

Many US sawmills sell their lumber on an *FOB mill* basis. FOB stands for *free on board* and means that the mill is selling its lumber loaded onto the conveyance chosen to ship it--be that rail, truck, or van. Implied in the term FOB mill is the idea that the sawmill is simply selling the lumber and loading it out. The mill is in no way involved with the freight to destination, once the car or truck is loaded and the proper paperwork has been completed. When buying from a mill that sells this way, the freight bill and all freight considerations are strictly for the buyer's account.

Some sawmills, on the other hand, will **quote** a mill-sales price for their lumber, but will only **sell** their lumber on a *delivered price* basis. When buying on delivered price terms, the lumber is bought delivered to the destination with a price that includes the freight. Now, in most cases, lumber being bought has not yet been shipped. Therefore, it has not been weighed and so the freight bill has not been written. A mill that sells on a delivered price basis, then, must calculate freight based upon an estimate of what the lumber will weigh. To determine the delivered price, the mill takes the estimated weight per MBF of the lumber, multiplies times the freight rate, adds this freight figure to the mill price of the lumber, and arrives at a *delivered to destination* price. The whole thing is quoted in terms of a single price per MBF.

For example, let's say that the freight rate from Prince George, British Columbia, to Miami, Florida is 4.98 on 85 ($4.98/cwt on 85,000 pounds minimum). The mill, located in Prince George, sells its lumber on a delivered price basis and uses an estimated weight of 1800 pounds per MBF. (1800 lbs/MBF is a commonly used estimated weight for much of the dry lumber produced in Canada.) To calculate a delivered price per MBF then, the mill will take the freight rate multiplied times 18 (the number of hundred-weights/MBF), and add it to the mill price of the lumber. Let's suppose that the mill price of the lumber is $200/MBF. The mills delivered selling price, then, will be $4.98 x 18 = $89.64 + $200 = $289.75/MBF.

Just because a mill is selling its lumber on a delivered price basis does not necessarily mean that it is going to pay the freight bill. There are two ways that freight bills in these cases can be handled: either the mill can pay the freight bill or the trader can pay the freight bill. If the mill pays the freight bill, it will send out it's invoice showing the entire delivered price, multiplied times the footage shipped, and the total due.

If the mill sells on a delivered price basis but wants the trader to pay the freight bill, the mill's invoice will show the delivered price, multiplied times the footage shipped, then allow a deduction equal to the actual. The actual freight is invoiced from and paid to the railroad company by the trader's company.

There are reasons why mills sell their lumber on a delivered price basis, and they have to do with who pays the freight bill. In the first instance, where the mill pays the freight bill, the advantage for the mill is in negotiated freight rates. That is, if a mill can negotiate special contract rates with certain rail carriers (rates which are cheaper than the published rates everyone else must use), and if the mill then keeps itself as the shipper on all rail shipments, the mill can pay the freight invoice on the basis of the cheaper, negotiated rate, yet charge the customer on the basis of the higher published rate. The difference is added profit for the mill.

In the second instance, when the trader pays the freight bill then deducts the actual freight charge from the mill's invoice, the advantage for the mills is the *underweights*. Underweights are the difference between the actual weight of the lumber and the weight that the mill uses when calculating the delivered price of its lumber.

The estimated weights/MBF that a mill uses to determine its delivered pricing is usually higher than the actual weight/MBF of the lumber. For example, let's say that a mill is selling lumber into a $5.76/cwt freight-rate area. The mill says that it is going to use estimated the weights of 1800 (read as "18 hundred") pounds per MBF. (These estimated weights are sometimes referred to as *association weights*.) The freight on this sale, then, will be assessed at 18 x 5.76, or $103.75/MBF. Now, both the trader and the mill may know that the lumber will only actually weight 1650 pounds per MBF. Therefore, they both know that the mill is charging for 150 pounds per MBF that do not exist. This 150 pounds is called the mill's "underweights." On a car shipping to a 5.76 rate area, a mill that gets 150 pounds in underweights receives additional revenue of $5.76 x 1.5, or $8.75/MBF. Therefore, a car loaded with 70 MBF represents $656.25 additional revenue for the sawmill.

Underweights may seem to be an unfair charge. However, because mills that sell and keep the underweights must compete with mills that don't (that is, mills that sell on an FOB mill basis), what usually happens is sawmills that sell and keep the underweights must lower their mill asking price for the lumber. In this way, the extra revenue generated with the underweights is given back in the form of a cheaper mills sales price for the lumber. It is primarily Canadian sawmills that sell their lumber on the basis of underweights. Why do mills fiddle with underweights at all? In this day and age, its simply marketing strategy.

A second peculiarity in buying Canadian lumber is something called *the prepay*. The prepay refers to when and how freight bills are paid on shipments of lumber from Canada to the United States.

Several years ago, most freight bills were paid upon arrival at the destination, or freight collect. After all, the amount of the bill was the same whether it was paid in advance or paid upon arrival. Rather than prepaying the freight, shippers knew that they could keep their freight dollars in the bank earning interest until the lumber reached its destination. Not surprisingly, this is what they did.

The railroads, for the same reason, and particularly the Canadian railroads where the transit time tended to be long, decided that they would try to get their freight money in advance.

All payments by United States shippers to Canadian carriers is made in Canadian funds, and so requires the converting of US currency into Canadian currency at the current bank exchange rate. To encourage payment in advance (meaning as soon as the car is loaded, weighed, and the freight bill is made out), the Canadian carriers started giving an exchange rate better than the bank's exchange rate, to shippers who would prepay the freight. The difference between the actual exchange rate and the railroad's preferential exchange rate is called the prepay. The exact preferential rate is different with each Canadian carrier, but it typically amounts to $5-$10/MBF, depending upon the originating carrier, the destination, the weight of the lumber, and the current bank exchange rate.

U.S. carriers, who do not have the exchange rate option to encourage prepayment of freight, usually make prepayment of freight bills a mandatory part of special freight contracts or incentives that they offer to shippers.

Like underweights, the prepay is one more factor in determining the net delivered price/MBF for a carload of lumber. Keep in mind, though, that the real variable in this equation is the mill's sales price; the determining issue is the net delivered price. All mills must compete.

Still one more factor in the freight-weight game is the growing practice of *agreed weights*. In the past, a loaded railcar had to be physically weighed, or *scaled*, before a freight bill could be issued. Weighing cars is expensive, time consuming, and means that freight bills cannot be issued as quickly as they might be. To overcome these obstacles, some railroads have begun a practice of weighing a random sample of cars, each containing the same lumber item and coming from the same mill. With this information, the railroad can determine an average weight/MBF for each item from each mill. The railroads then use these average sample weights/MBF for billing purposes, instead of actually weighing the cars. These random sampling averaged weights are called *agreed weights.* A mill shipping on agreed weights knows what its agreed weights are. In this way, the mill can tell precisely what weight a car with a certain item and amount of footage will be billed on, even before it is loaded.

Calculating weights in cars would be more exact but for two facts. The first is that sawmills do not always get exactly the car size that they order from the railroad, and each car size loads different footage. The second is that mills do not always know exactly what lumber footage will be available for loading until an order is ready to ship.

Agreed weights and average (or association) weights do, however, make it easier to predict freight charges/MBF before lumber is loaded. To illustrate, how agreed weights work, let's go back to the first example of a railcar shipping lumber from Spokane, Washington to Chicago, Illinois. The rate for that haul was 3.45 on 85, 3.20 on the excess. Let's say that we don't know what car size the mill will get from the railroad for shipping this order, but we do know that the agreed weight of the lumber is 1675 lbs/MBF. We also know that, as most mills will, this mill has agreed to load enough lumber to at least surpass the 85,000 pound minimum weight. Taking this information, we can estimate the freight charges closely by taking the *first factor*, or first part of the rate (3.45 on 85 in this case), and multiplying times the agreed weight, as follows: $3.45 x 16.75/MBF = $57.75/MBF.

By adding $57.75/MBF to the mill price of the lumber, we have a close estimate of the net cost delivered to the destination.

One last example: the 110,000-pound boxcar rate from Prince George, British Columbia to Savannah, Georgia is 5.17/cwt. The mill is going to sell on average weights of 1800 lbs/MBF. To calculate a delivered price, multiply the rate by 18 as follows: $5.17 x 18 = $93.00

Add $93.00/MBF to the mill price per MBF to arrive at the actual cost delivered per MBF price. If we choose to prepay the freight bill, we can take advantage of the preferential exchange rate and reduce the delivered cost by $5-$10/MBF.

Because railroad traffic has become so specialized in the last few years, most trading companies employ people who are experts in the area of rates and routing. If a company does not provide people specialized in this area, the information is available by phoning the originating or delivering carrier. That carrier will work out a routing and quote an exact rate from the origin to the destination, although the routing they use may not necessarily be the cheapest. (The routing you are liable to get is the one that gives that carrier the most favorable haul.)

TRUCKING

Most over-the-road flat-deck trucks have from 40' to 48' of loadable space, they are approximately 8' wide, and they can carry approximately 48,000 pounds of lumber.

Truck rates are usually quoted one of two ways. The first is on the basis of the number of *loaded miles*. This means that the rate is calculated at a fixed price per mile on the actual number of road miles from the origin to the destination, regardless of the amount of lumber being carried. For example, to haul a load of lumber from Columbia Falls, Montana to Milwaukee, Wisconsin may cost $1.10 per loaded mile. If the distance between Columbia Falls and Milwaukee is 1,500 miles, the freight bill for such a haul will be 1,500 x $1.10 = $1,650.

Again, because lumber is quoted in terms of a delivered price per MBF, the trader must convert this dollar amount of $1,650 into a "per MBF" price. To do this, the approximate footage, or number of MBF, a truck will carry must be known. Let us take an example where the lumber being carried weighs approximately 1,700 pounds per MBF. (Lumber weight varies by size, species, moisture content, and surfacing. The mill can give you a close approximation of what lumber will weigh.) You know that the truck can carry approximately 48,000 pounds. (The maximum weight allowance of a truck includes the weight of the tractor and the trailer itself, so the amount of cargo that can be hauled varies directly with the weight of the truck and trailer hauling it. A light truck and trailer rig may be able to haul as much as 3 or 4 MBF more than a heavy truck and trailer rig.) By dividing 48,000 pounds by the weight per MBF (1,700), you can calculate that the truck should be able to carry between 28 and 28.5 MBF of lumber on this load. So, by dividing the total freight charge ($1,650) by the number of MBF (28), the freight price per MBF can be calculated: $1,650/28 MBF = $59.00/MBF.

This figure of $59.00/MBF is a very close approximation of what the actual freight charge will be. Until the truck is actually loaded, though, you won't know precisely how much footage has been put on, and, therefore, what the exact price per MBF will be. (Trucks are not loaded while resting on a scale, so the person who loads the truck at the mill will estimate the weight being loaded to the nearest whole, or sometimes half, unit of lumber.)

The second way that truck rates are quoted is by a price per hundredweight. This is basically the same type of quotation as most railroad rate quotations. For example, let's suppose the rate from Columbia Falls, Montana to Milwaukee, Wisconsin is quoted on the basis of a price-per-cwt rate. The quoted rate might be 5.78/cwt on a minimum of 48,000 pounds. As with rail rates, this type of truck rate includes a minimum weight that will be charged, regardless of the weight that is actually carried. If we again assume that the lumber weighs 1,700 pounds per MBF, we can calculate the delivered freight price using either of two methods.

One method is to take the rate ($3.47/cwt), times the weight/MBF (1,700 pounds/MBF) and multiplying: $3.47 x 1700/MBF = $59.00/MBF.

Another method for calculating the rate is by taking the truck's minimum weight of 48,000 pounds (or, 480 hundredweights), times the rate ($3.47/cwt) and dividing by the total footage being shipped: ($3.47 x 480)/28 MBF = $59.50/MBF.

You will use the first method of calculation when you are sure that the mill can, and will, load up to or over the minimum weight. Use the second method if you know exactly how much lumber the mill will load on the truck. Using either method, the freight figure is a close approximation of what the actual freight bill will be.

There is one key difference between truck rates quoted as a price per cwt and rail rates quoted as a price per cwt. That difference is that rail cars have very high maximum weights--much higher than the minimum weight upon which the rate is based. This encourages the loading of much more lumber per car than the minimum weights would indicate. Trucks, on the other hand, have very low maximum weights. Truck maximum weights are frequently at, or only slightly higher, than the minimum on which the rate is based. For example, a truck rate may be based on a minimum of 48,000 pounds, yet have a maximum of 48,500 or 49,000 pounds.

PIGGYBACK VANS

Pig vans, as they're referred to, are truck-trailer vans completely enclosed on the front, the sides, the top, and the bottom. They are open only at the rear. They are designed for travel both over the road like conventional trucks, or secured onto railroad flatcars and transported by rail.

Piggyback van rates are different than either truck or rail rates. Piggyback vans come in different width sizes, each of which has a different rate. The most commonly used van size for lumber is 102 inches wide. Because they do get hauled over the road, pig vans are subject to the same weight restrictions as normal trucks. For this reason, they carry just about the same volume and weight as a normal truck.

Piggyback van rates are usually quoted as a flat charge per trip. For example, the basic rate from Spokane, Washington to Columbia, South Carolina might be $2,708. This is called a *door-to-door* rate. Door-to-door service means that the pig van is loaded at the mill, by the mill, and is carried over the road by a regular truck tractor to the nearest railroad loading ramp. There it is rolled on, and secured to, a railroad flatcar (two vans per flatcar). The van, on the flatcar, is then delivered as close as possible to the destination city by rail. Upon arrival, it is rolled off the railcar, again attached to a truck tractor, and delivered to the customer. At the destination, the customer unloads the van, and it is hauled away.

Because vans are loaded only from the rear, they require special loading skills and facilities. Not all mills can or will load pig vans. In a case where the mill doesn't load vans, they can still be used. However, in such an instance, the lumber must be picked up at the mill on a normal flatdeck truck, hauled to a reload center equipped to load (or *stuff*) vans, and then delivered to the destination in the same fashion described above for door-to-door service. The type of rate for this service is called a *flat-to-van* rate. Because of the extra handling on the loading end, this rate is always higher than the door-to-door rate.

Occasionally the reverse of flat-to-van service is required. That is, the mill can load the pig van but the customer cannot unload it. In this case, a *van-to-flat* rate is needed. A van-to-flat rate means that the lumber is stuffed in the van at the mill, travels to the nearest destination city by rail, and is off-loaded from the railcar. The lumber is then unloaded from the van and placed on a flat-deck truck for delivery over-the-road to the customer's yard. Because of the extra handling on the delivering end, this rate is also higher than the door-to-door rate.

It is possible to get a flat-to-van-to-flat rate for a situation in which the mill doesn't load and customer doesn't unload vans. However, because of all the extra handling, the cost is usually prohibitive. In such a situation, a fully tarped over-the-road truck is usually less expensive and, because it's not handled as much, the lumber will probably arrive in better shape.

Sometimes a mill can load its lumber into vans right at the loading ramp where the railroad loads the vans onto the railcars. The rate needed for this loading is called *ramp-to-door* (or a *ramp-to-ramp* if the customer can pick up the van at the destination ramp with a truck). This rate is cheaper than a door-to-door rate because it does not involve any over-the-road trucking.

Because vans do require extra care in loading, and because they do not load as quickly as a truck, mills that load vans will frequently add a surcharge of $5-7/MBF for this extra service.

Most pig vans are handled by special freight agents. These agents take care of the reloading of the material, if necessary; and are responsible for the railroad routing and freight bills. They then include these costs in the quoted van price.

BARGING

As mentioned in the introduction to this chapter, a fourth form of transportation for lumber is ocean-going barges. Ocean-going barges carry huge quantities (many millions of board feet) of commodity-type lumber (like 2x4, 2x6, and studs) through the Panama Canal to eastern seaboard docks. The time frame is quite long. It typically takes two to four months from the time that lumber is bought at a mill until it arrives at the east coast dock. Because of this, barge freight is highly specialized and generally is not a consideration in day-to-day lumber trading.

CHAPTER
FOUR

Covering the Basics, 20 Key Questions

Every lumber trade involves two separate and distinct orders: the order buying lumber from the sawmill, the buy order, and the order selling lumber to the customer, the sell order. These two orders **must** agree. All parties, meaning the sawmill, the buying trader, the selling trader, and the customer, must understand and concur on the particulars of these two orders.

To ensure agreement on the orders, there are twenty key issues that must be addressed and reconciled before a trade is finalized. A misunderstanding on any of these issues can lead to costly and time-consuming claims, as well as mill and customer dissatisfaction. What follows is a close look at each of these twenty key issues, posed in the form of questions.

THE TWENTY QUESTIONS OF LUMBER TRADING

Characteristics of the Lumber

1. What is the species?
2. What is the dimension (thickness and width)?
3. What is the grade?
4. What is the moisture content?
5. What is the surfacing?
6. Are the edges eased?
7. What is the end trimming?
8. Is the lumber gradestamped?

Packaging of the Lumber

9. How many pieces per unit?
10. How many lengths per unit?
11. How is the lumber protected?

What, When, and How Is the Lumber To Be Shipped?

12. What is the approximate quantity?
13. What is the tally?
14. What is the loading?
15. What is the shipment time?
16. What is the customer's delivering carrier or truck delivery address?

Pricing and Terms of Payment

17. What is the price?
18. What is the customer's credit status?
19. What are the terms of payment?
20. What is the customer's order number?

An explanation of each of these questions follows. For simplicity, the discussions will be from the point of view of the trader as the seller. The same questions and explanations apply, but in reverse, when the trader is the buyer.

CHARACTERISTICS OF THE LUMBER

1. What Is the Species?

Because each species of wood contains its own unique characteristics of strength, texture, knot structure, appearance, weatherability, etc., the species being traded must be clearly identified. Listed in Table I are the major commercially-traded softwood species of western North America.

The primary uses of each species of wood has been determined over the years by its characteristics and availability. For the most part, each species of lumber is manufactured and marketed separately. However, it is industry practice that when certain different species grow and are harvested together, and when those species have similar properties (strength, appearance, etc.) that make them interchangeable in use, they may be manufactured and marketed together as a *mixed-species grouping*. The common mixed-species groupings for western lumber are listed in Table II. When looking at Table II, the word in parentheses next to the group name is the shorthand form of the group name that appears on an actual gradestamp. Also, as some species groupings are strictly Canadian, some are strictly United States, and some are both, this information appears below the grouping name.

TABLE I

THE MAJOR COMMERCIALLY-TRADED SOFTWOOD SPECIES OF WESTERN NORTH AMERICA

Group	Common Name	Botanical Name
True Firs		
	Alpine/Subalpine	(Abies lasiocarpa)
	Balsam	(Abies balsamea)
	California Red	(Abies magnifica)
	Grand	(Abies grandis)
	Noble	(Abies procera)
	Pacific Silver	(Abies amabilis)
	White	(Abies concolor)
Spruce		
	Engelmann	(Picea engelmannii)
	White	(Picea glauca)
	Sitka	(Picea sitchensis)
Pine		
	Idaho White	(Pinus monticola)
	Ponderosa	(Pinus ponderosa)
	Sugar	(Pinus lambertiana)
	Lodgepole	(Pinus contorta)
Hemlock		
	Mountain	(Tsuga mertensiana)
	Western	(Tsuga heterophylla)
Cedar		
	Alaska Yellow	(Chamaecyparis nootkatensis)
	Incense	(Libocedrus decurrens)
	Port Orford	(Chamaecyparis lawsonia)
	Western Red	(Thuja plicata)
Other		
	Douglas Fir	(Pseudotsuga menziesii)
	Western Larch	(Larix occidentalis)
	Redwood	(Sequoia sempervirens)

TABLE II

COMMON MIXED SPECIES GROUPINGS FOR WESTERN LUMBER

(Listed Alphabetically)

GROUP NAME (Abbreviation) (Where manufactured)	SPECIES
Douglas Fir/ Larch (DF/L) (U.S. & Canadian)	Douglas Fir Western Larch
ES/LP or ES/LP/AF (U.S. only)	Engelmann Spruce Lodgepole Pine Alpine Fir (if ES/LP/AF)
Hem/Fir (Hem/Fir or H/F) (U.S. & Canadian)	Western Hemlock All true firs (except Alpine and Balsam)
Mixed Species (M/S) (U.S. only)	All western softwoods
Northern Species (Northern Species) (Canadian only)	All softwoods
PP-LP (PP-LP) (U.S. only)	Ponderosa Pine Lodgepole Pine
PP-SP (PP-SP) (U.S. only)	Ponderosa Pine Sugar Pine
SPF (SPF) (Canadian only)	Western White Spruce Engelman Spruce Black Spruce Red Spruce Lodgepole Pine Jack Pine Alpine Fir Balsam Fir

(continued next page)

(Table II, continued)

West Coast Woods (WCW) (U.S. only)	Douglas Fir Western Larch All true firs All true spruce All true hemlock
Western Cedar (WC) (U.S. & Canadian)	Incense Cedar Western Red Cedar Port Orford Cedar Alaska Yellow Cedar
Western Woods (Western Woods) (U.S. only)	Any combination of western softwoods (except redwood)
White Woods (WW) (U.S. only)	Engelmann Spruce All true firs All hemlocks All pines

2. What Is the Dimension (Thickness and Width)?

The word *dimension* has two distinct meanings in everyday lumber jargon. The first meaning is the nominal thickness and width measurement, usually in inches, of any piece of lumber. For example, 1x6 or 4x12 are the dimensions of particular sizes of wood. In these descriptions, the first number represents the thickness, the second number represents the width. In this use, the word *dimension* is synonymous with *size*.

The second way that the term dimension is used is when discussing lumber that is 2" thick. Lumber such as 2x4, 2x6,...2x12 is called dimension lumber. (Technically, dimension lumber includes everything from 2" to 4" in thickness. However, in practice, dimension lumber refers to that which is just 2" thick.)

As shown in Table III, lumber is generally classified into four primary types. These types are based upon **thickness.**

TABLE III

CLASSIFICATION OF LUMBER BASED ON THICKNESS

TYPE	NOMINAL THICKNESS
Boards*	1"
Dimension**	2" to 4"
Timbers	5" and thicker
Factory	4/4" to 16/4" (read four-quarter to sixteen-quarter)

* Board grades may be used on lumber up to 4" thick, but generally refer only to lumber that is 1" (nominally) thick.

** As noted, dimension lumber usually refers to lumber 2" in thickness.

The majority of all softwood lumber is produced in widths, in 2" increments, from 4" to 12" (i.e., 4", 6", 8", 10", and 12"). A small amount of lumber is produced in widths of 2", 3", and odd sizes up to 24".

It is important to note that what we call a 2x4 or a 1x8 is usually a piece of lumber smaller than its *nominal* (name) size. A dry, surfaced 2x4, for example, actually measures only 1-1/2" thick and 3-1/2" wide. A dry, surfaced, 1x8 actually measures 3/4" x 7-1/4". Lumber is bought, sold, and invoiced based upon the nominal (or name) size (e.g., 2"x4"), not the actual size (e.g., 1-1/2" x 3-1/2").

Listed in Table IV are the standard nominal and actual sizes for dry and green fully-surfaced lumber.

TABLE IV

STANDARD NOMINAL AND ACTUAL SIZES FOR DRY AND GREEN SURFACED LUMBER

BOARDS

NOMINAL	NET DRY	NET GREEN
1x2	3/4" x 1-1/2"	25/32" x 1-9/16"
1x3	3/4" x 2-1/2"	25/32" x 2-9/16"
1x4	3/4" x 3-1/2"	25/32" x 3-9/16"
1x6	3/4" x 5-1/2"	25/32" x 5-5/8"
1x8	3/4" x 7-1/4"	25/32" x 7-1/2"
1x10	3/4" x 9-1/4"	25/32" x 9-1/2"
1x12	3/4" x 11-1/4"	25/32" x 11-1/2"

DIMENSION

NOMINAL	NET DRY	NET GREEN
2x2	1-1/2" x 1-1/2"	1-9/16" x 1-9/16"
2x3	1-1/2" x 2-1/2"	1-9/16" x 2-9/16"
2x4	1-1/2" x 3-1/2"	1-9/16" x 3-9/16"
2x6	1-1/2" x 5-1/2"	1-9/16" x 5-5/8"
2x8	1-1/2" x 7-1/4"	1-9/16" x 7-1/2"
2x10	1-1/2" x 9-1/4"	1-9/16" x 9-1/2"
2X12	1-1/2" x 11-1/4"	1-9/16" x 11-1/2"

FACTORY
(Dry Only)

NOMINAL THICKNESS	STANDARD THICKNESS	OPTIONAL THICKNESS
4/4	3/4"	25/32", 7/8"
5/4	1-5/32"	1-1/4", 1"
6/4	1-13/32"	1-9/32", 1-1/2"
8/4	1-13/16"	1-7/8", 1-15/16"
10/4	2-3/8"	
12/4	2-3/4"	
16/4	3-3/4"	

TIMBERS
(Green Only)

NOMINAL THICKNESS AND WIDTH	STANDARD SURFACED THICKNESS AND WIDTH	ROUGH (STANDARD SAWN)
5x5 & lgr	1/2" off nominal	not more than 3/8" under the nominal size

3. What Is the Grade?

Lumber is graded so as to sort it into different types of use categories, and to establish a uniform standard between different lumber manufacturers. This is done so that lumber of a given grade will have the same value and can be used for similar purposes, regardless of the producing region or the type of log from which it was taken.

Lumber grades are based upon the cumulative effect, on a piece of lumber, of the natural characteristics of the wood, the quality of the manufacturing, and any seasoning effects (i.e. like improperly drying). Each grade defines the limiting characteristics that a piece of lumber may contain, yet still be classified within a particular category. Some grades are based primarily upon appearance properties; some are based primarily upon structural and strength properties; and some incorporate both appearance and strength properties.

Over the years, voluntary regional manufacturers associations, working in conjunction with the American Lumber Standards Committee and/or the Canadian Lumber Standards Accreditation Board, have developed voluntary product standards for all types of lumber items. These associations, among other things, establish and oversee standardization and compliance to grade by their member mills. Each grading association publishes a grade book describing various grades that their mills may designate. Although grade characteristics are similar between the major grading associations, a trader should keep a current copy of, and become familiar with, the necessary grade rules for each of the different associations. In the case of a disagreement on grade, it is the mill's association's rules that will be used as the criteria on which to judge the lumber.

Each individual piece of lumber coming from a sawmill is given a grade. Most lumber is graded visually, piece by piece. Boards, because they are normally used in appearance applications, are graded upon the basis of the face (best wide side.) Factory lumber is usually graded on the basis of the poorer of the two wide sides. Dimension lumber and timbers, because they are primarily used in structural applications, are graded for strength: all four sides, therefore, are considered in the grade.

What follows now is a look at the most commonly traded grades, by lumber type, of western softwood lumber. This list is by no means all-inclusive, as there are many specialty grades written for specific applications. The list does, though, cover probably 98% of all western lumber traded domestically.

Table V-A describes normal board grades. There are two generally employed classifications of board grades. The first, and most widely used, is called "common" rules and is generally used for grading the inland species (i.e. spruce, pine and true firs.) The second classification , sometimes referred to as "alternate rules" is applied primarily to coastal species (Douglas Fir, Western Hemlock, most cedars, and Sitka Spruce.) As will be seen, there is some overlap between the groups.

It should also be noted that although most grading association distinguish between "common" grades and "select" grades, "selects" are generally traded as though they are the upper end of the common grades. For this reason, they have been included in with the board grades and are not dealt with as a separate group.

Table V-B lists the dimension grades. Different than the other three types of lumber, all softwood **dimension** lumber 2" to 4" thick graded under the auspices of accredited agencies **must** conform to the National Grading Rule for Softwood Dimension Lumber. These rules are published by the US Department of Commerce and are administered by the American Lumber Standards Board of Review, in conjunction with the Canadian Lumber Standards Accreditation Board.

TABLE V - A

MOST COMMONLY TRADED BOARD GRADES, BY SPECIES, OF WESTERN SOFTWOOD LUMBER

BOARD GRADES (1" THICK)
(Idaho White Pine (IWP) has different grade names for the same grades as other commons. The IWP grade names are given in parentheses.)

Grades: Commons	Species
C & Better Select (Choice & Better IWP)	Ponderosa Pine, Sugar Pine Idaho White Pine
D & Better Select	ES/LP, SPF, Doug Fir/Larch, western cedars
D Select (Quality IWP)	Ponderosa Pine, Sugar Pine, Idaho White Pine
2 & Btr Common (Sterling IWP)	All pine & spruce
3 & Better Common	Doug Fir/Larch, western cedars
3 Common (Standard IWP)	All pine & spruce
4 & Better Common	ES/LP, SPF
4 Common (Utility IWP)	All pine, spruce & western cedars
5 Common (Industrial IWP)	All species

Grades: Alternate Board Grades

Species: Doug Fir, Fir/Larch, Hem/Fir, Sitka Spruce, all western cedars

Superior Finish
Prime Finish
E Finish
Select Merchantable
Construction
Standard
Utility
Economy

SPECIAL WESTERN RED CEDAR GRADES

Grades: Finish, Paneling, and Ceiling

Clear Heart
A
B

Grades: Tight Knot Paneling & Siding

Select Knotty
Quality Knotty

Grades: Bevel Siding

Clear VG Heart
A Bevel
B Bevel
C Bevel

REDWOOD GRADES

Grades:

Clear All Heart
Clear
Heart B
B
Select Knotty
Construction Heart
Construction Common
Heart Merch
Merch
Economy

TABLE V - B

MOST COMMONLY TRADED DIMENSION GRADES, BY SPECIES, OF WESTERN SOFTWOOD LUMBER

DIMENSION GRADES (2" - 4" THICK)

SPECIES: All Western Softwoods

Grades: Light Framing Rules (Lumber 2"-4" Thick, 2"-4" Wide)

Construction
Standard
Utility
Economy

<u>Grades: Structural Light Framing</u> (Lumber 2" - 4" Thick, 2" - 4" Wide)

Dense Select Structural (Doug Fir only)
Select Structural
Dense #1 (Doug Fir only)
#1
Dense #2 (Doug Fir only)
#2
#3
Economy

<u>Grades: Stud</u> (Lumber 2"-4" Thick, 2"-6" Wide, 10' & Shorter)

Stud
Economy Stud

<u>Grades: Appearance Framing</u> (Lumber 2"-4" Thick, 2" & Wider)

Appearance

<u>Grades: Structural Joists and Planks</u> (Lumber 2"-4" Thick, 5" & Wider)

Dense Select Structural (D Fir only)
Select Structural
Dense #1 (Doug Fir only)
#1
Dense #2 (Doug Fir only)
#2
#3
Economy

Some lumber, in addition to being graded visually, is also mechanically tested for stiffness and strength. This process is called machine stress rating (MSR). To determine its MSR rating, each piece of lumber is non-destructively pressure tested and assigned a strength figure, based upon the test result. This figure is then included in the gradestamp and marked on the lumber. MSR testing is normally just done on dimension lumber, and occasionally on boards.

Listed in Table V C are the most commonly-traded factory grades of western softwood lumber. It is in the area of factory lumber that there are the most numerous highly specialized grades. Consult individual association's gradebooks for full listing of these specialty products.

TABLE V - C

MOST COMMONLY TRADED FACTORY GRADES, BY SPECIES, OF WESTERN SOFTWOOD LUMBER

FACTORY (4/4"-16/4" THICK)

SPECIES: Ponderosa Pine, Sugar Pine, Douglas Fir, Hemlock, and White Fir

Grades: Moulding and Shop Lumber

Moulding & Btr
Moulding
3rd Clear
#1 Shop
#2 Shop
#3 shop

Listed in Table V-D are timber grades.

TABLE V - D

MOST COMMONLY TRADED TIMBER GRADES, BY SPECIES, OF WESTERN SOFTWOOD LUMBER

TIMBERS (5" AND THICKER)

SPECIES: Douglas Fir, Fir/Larch, Hem/Fir

Grades: Beams and Stringers
(5" & Wider, Width more than 2" greater than thickness)

Select Structural
#1 (#1 Structural)
(#2 Structural)
#2 (Standard)
#3 (Utility)

Grades: Posts and Timbers
(5" & Wider, Width not more than 2" greater than thickness)

Select Structural
#1 (#1 Structural)
(#2 Structural)
#2 (Standard)
#3 (Utility)

4. What Is the Moisture Content?

Moisture content is a critical issue that must be discussed any time lumber is bought or sold. Moisture content affects lumber in the following ways:

1. The lower the moisture content, the lighter the lumber. Since transportation charges are usually

based upon weight, lower moisture content will usually mean lower freight charges.
2. A lower moisture content means that lumber is less likely to develop problems from stain, decay, or insects.
3. A lower moisture content will help prevent shrinkage and *checking* (end splitting) after lumber is in use.
4. A lower moisture content makes wood more stable; more inclined to stay straight and flat.
5. A lower moisture content increases strength and stiffness.

The moisture content of a piece of wood is defined in most gradebooks as the weight of the water in the wood, expressed as a percentage of the weight of the oven-dry wood. (Oven-dry wood is wood from which all the moisture has been removed.) This means that lumber with a moisture content of 15% contains 15 pounds of water for every 100 pounds of oven-dry wood fiber.

Because the drying process is time consuming and an additional expense, mills dry their lumber only when they feel that the sales price will justify the additional cost. For this reason, most construction lumber that is used close to the lumber-producing region (where transportation costs make up a small part of the total price) is not dried. On the other hand, construction lumber shipping great distances (with corresponding high freight) usually is dried.

It should also be noted that thick lumber, over 2", is seldom dried. In order for it to remain stable, lumber must have approximately the same moisture content throughout its thickness. Lumber over 2" thick is difficult to dry uniformly. It takes a long time and therefore is quite expensive.

Most lumber that is dried is put through a *drying kiln* in which both moisture and temperature are controlled. Due to the cost of building and maintaining a drying kiln, some smaller sawmills do without them. Instead, these smaller mills rely on air drying for any dry lumber that they produce.

Air drying lumber means putting the units of rough, stickered lumber outside, exposed completely to the elements. As air moves through the units, moisture is evaporated away.

Although there are advantages and disadvantages to both air drying and kiln drying, if done properly the finished product is basically the same for most purposes. Most gradestamps that indicate moisture content do not distinguish between lumber that has been kiln dried and lumber that has been air dried. Gradestamps generally only indicate that the lumber is dry.

To describe the moisture content for any particular lumber, the following terminology is used:

Green (or unseasoned). Green is the term used to describe the moisture content of lumber just as it comes out of the log. Nothing has been done to reduce its moisture content. Green lumber will, over time, lose moisture naturally until it reaches equilibrium with the environment into which it is placed. (i.e., it will become air dried).

Dry (either kiln-dried or air dried). Dry means that lumber contains 19% or less moisture content. "Dry" select and finish grades of boards, and factory lumber, must be dried to 15% or less moisture content. Most lumber shipping beyond the producing region, as well as most board and factory items, are dried

Moisture Content 15 (MC 15). Lumber that has been gradestamped MC 15 must contain 15% or less moisture content. This extra dryness ensures greater stability and is seen most frequently in selects, factory, or lumber to be used in furniture or manufacturing applications.

Partially Air Dried (PAD). As the name implies, PAD stock is not dry, and yet it is not absolutely green either. If lumber is truly PAD, it is usually material that has been allowed to air dry for some time before being sold. Great care must be taken any time PAD material is traded. This care must be exercised because there is no guarantee of the lumber's moisture content. Unless some verifiable information about moisture content can be obtained, PAD lumber should be treated as though it were absolutely green.

Kiln Wet. Kiln-wet lumber is lumber that has gone through the dry kiln, but for some reason its moisture content has not been reduced to the 19% required for dry lumber. Mills do not try to produce kiln-wet lumber. The most common cause for kiln-wet lumber is mixed species in the kiln: different species do not always dry at the same rate.

Kiln-wet lumber is sometimes implied, by a seller, to be close to 19% moisture content. As with PAD lumber, though, because no moisture guarantee is made, great caution should be taken. If kiln-wet lumber is gradestamped, it must be gradestamped "green", not "dry." However, because at most mills lumber goes through the planer and past the grader before it is finally checked for moisture content, the kiln-wet pieces will usually already be surfaced to dry sizes. This means that the kiln wet lumber, as it air drys, will shrink to slightly smaller than normal dry sizes.

5. What Is the Surfacing?

Surfacing makes the sides of a piece of lumber smooth and uniform in thickness. Most commercially-used board and dimension lumber is planed (surfaced, also referred to as *dressed*) on all four sides. This surfacing is referred to as *S4S*, meaning *surfaced four sides*.

Factory lumber is sometimes surfaced S4S, other times it is not. Most factory lumber is purchased by manufacturers such as door, window, and furniture makers, who intend to rework the lumber into smaller pieces. Smooth and uniform edges are not as important to these customers as is slight extra width. With extra width, they can frequently increase the number of cuttings they get from the lumber. Therefore, factory lumber is sometimes sold surfaced only on the wide faces. This surfacing is called *S2S*, which means *surfaced two sides*. The edges (the narrow faces) are not surfaced on S2S lumber. In some cases, the edges on factory lumber are not even trimmed. This leaves boards that may be wider at one end than the other. Again, this extra width can be utilized by a factory or millwork plant.

Occasionally, a piece of S4S lumber will be remanufactured, at the sawmill, into another product. This remanufacturing is done by either *resawing* or *ripping* the material, both of which change the physical size of the lumber, or running a *pattern* on the lumber, which changes its S4S appearance.

Resawing is the process of cutting lumber in two, usually through the middle, parallel to the wide face. In other words, resawing changes the **thickness** of the lumber but not its width.

For example, if a piece of 2x12 is resawn, the result will be two pieces of 1x12. These 1x12s, however, will be slightly thinner than normal S4S 1x12. Normal 1x12s are 3/4" thick. The two newly-made 1x12s will each be approximately 11/16" thick. The difference, or loss, in thickness of the resawn material is due to the 1/8" saw *kerf* (the gap created by the saw blade).

Since the piece of 2x12 being resawn starts out S4S, after resawing each piece will be surfaced on one side (the outside wide face) and two edges (the narrow faces). It will not be surfaced on the inside face where the saw has just passed. This new face is rough. The name for the surfacing on this resawn lumber is S1S2E, or, *surfaced one side, two edges*.

Ripping is the process of changing the **width** of a piece of lumber by cutting it perpendicular to the wide face. A piece of S4S 2x12 might be ripped to produce two pieces of 2x6. As 2x12 is normally 11-1/4" wide, each of these new 2x6s will be approximately 5-11/16" wide. Normal S4S 2x6 is 5-1/2" wide.

Because the 2x12 prior to being ripped was surfaced S4S, the new pieces of 2x6 will be surfaced on the two wide faces and one edge. The new edge, on the inside just ripped, is rough. The name for the surfacing on this ripped lumber is *S2S1E*, or *surfaced two sides, one edge*.

Similarly, a 2x12 being *gang-ripped* into 2x2s will make six pieces of lumber. The outside two pieces will be surfaced S2S1E. (In the case of 2x2s, where all four sides have the same nominal width, the surfacing can be called S3S). The four pieces from the center will have a surfacing of simply S2S.

Pattern work is done by sending surfaced lumber back through the planer. Predetermined notches and grooves are inscribed in the lumber as it is rerun. These notches and grooves leave a pattern down the entire length of the board. Different patterns are used for different end uses. *Tongue and groove*, *channel rustic*, and *shiplap* are a few common patterns. Since patterns are put on after lumber has already been surfaced and graded, sawmills charge a *rerun*, or *pattern, fee* for providing this service. A typical pattern or rerun fee is between $35 -$50/MBF.

If a piece of lumber is not surfaced, it is called *rough*. Lumber that is sold in the rough will vary in sizing. There are, however, two terms used to describe the actual thickness and width of rough lumber. These terms are *full sawn* and *standard sawn*.

Full sawn means that lumber is cut in the sawmill to its full nominal size. That is, a 6x12 will actually measure 6" by 12".

Standard sawn means that lumber is cut slightly smaller than full sawn lumber. However, standard-sawn lumber is enough larger than its surfaced size, that when it is run through a planer, all four sides can be made smooth and uniform, to normal finished size.

One other surfacing option exists, and it is called *saw-sizing*. Saw-sized lumber is cut in the mill to finished surfaced sizes. It is not intended to go through the planer, yet can be used interchangeably with dressed lumber in applications where smooth surfacing is not important, or where a textured (rough) face is preferred.

6. Are the Edges Eased?

While being surfaced, the edges on some types of lumber are purposely rounded to reduce splintering while handling. These rounded edges are referred to as *eased edges*. On 1" lumber, the maximum rounding, or easing of the edges, is to a radius of 1/16". On 2" lumber, the maximum rounding is to a radius of 1/8".

Edges are generally eased only on dimension lumber, not on boards, timbers, or factory lumber.

If the edges on lumber are not eased, they are called either *not eased*, or *square edge*. *Not eased edge* means that the edges have not been purposely rounded. *Square edge* technically means that the edges are square and free of *wane* (wane is the absence of wood along an edge). Sometimes, incorrectly, the terms are used interchangeably.

7. What Is the End Trimming?

Most lumber--boards, dimension, timbers, and factory lumber--is *double end trimmed* (DET). This means that each end is cut reasonably square and smooth, and that the piece is at least as long as the invoiced length. On lumber 2" and thinner, overlength on DET lumber may be no more than 3".

A second type of end trimming is called *precision end trimmed* (PET). This is the name given to the end trimming on studs and on lumber with specific industrial applications, where requirements call for many pieces of lumber all cut to precisely the same length. Lumber represented as PET may vary only 1/16" in no more than 20% of the pieces.

Occasionally, a customer is **purposely** charged for smaller pieces of wood than is actually shipped. In other words, the customer gets more wood than is being invoiced (charged) for. This practice is referred to as *pencil trimming.* Pencil trimming is an invoicing procedure used by mills to sell unusual or hard-to-move items.

For example, a stud mill may have accumulated odd-trimmed studs between 60" and 72" in length. Because the mill does not have enough of any one trim size to make a full load, it might choose to sell them all *as is*. To do this, the mill will represent, sell, and invoice the material as *60" pencil trims*. The buyer of this lumber then knows that each piece is at least 60" long and that some pieces will be longer. Even on the longer pieces, however, the buyer pays for only 60".

By pencil trimming a sawmill can save the cost of physically rerunning lumber back through the trim saws or planer. This means that the mill can reduce the sales price, sell the lumber as is, and keep its inventory moving. The customer for pencil-trimmed lumber benefits by getting lower-priced lumber, as well as getting extra length. Pencil trimming is also done on width (i.e., 1x5 invoiced as 1x4, for example).

The customer must always approve of pencil-trimmed material before shipping, as this represents a change from the original order.

8. Is the Lumber Gradestamped?

Although all lumber is graded, not all lumber is gradestamped. As mentioned earlier, in order to assure uniform standards from mill to mill, grading associations have been established to define what defects and characteristics are allowable in each of the various grades, sizes, species, etc. When a lumber manufacturing company joins a grading association, each different facility (some companies operate numerous mills at different locations) is given an identification number. The mill, then, is authorized to put this number, in the form of an inked marking (a gradestamp), on the face of each piece of lumber that it produces.

In addition to the mill's identification number, the gradestamp includes the following information:

1. The name of the grading association to which the mill belongs (and, therefore, the rules under which the lumber is graded).
2. The species of the lumber.
3. The grade of that particular piece.
4. If the lumber was surfaced, whether it was surfaced green or dry.

The gradestamp may also indicate special moisture content (e.g., MC 15) or other pertinent information about how the lumber was produced.

Most dimension lumber, timbers, and some boards are gradestamped. Selects (highest grade boards), many middle- and low-grade boards, and factory lumber generally are not gradestamped. When lumber is not gradestamped, the sawmill will, upon request, provide a *certificate of grade* to go along with the shipment and verify its grade.

If a gradestamp is required on high-grade boards or selects, the mark may be put on the end of the piece. This practice is called *end-stamping.* With end-stamping a board's visual appearance is not reduced.

PACKAGING OF THE LUMBER

9. How Many Pieces Per Unit?

As a general rule, most lumber is packaged into units approximately 2' high and 4' wide. The number of pieces per unit varies with the size of the lumber and the size of the packages.

There has been a trend over the last several years toward larger-sized units. Larger units enable the loading of more board footage on railcars. By increasing the load on a car, the cost of freight per MBF is reduced. The sawmills and customers who make and take delivery utilizing these larger units can save on freight. Units much larger than 2' by 4', however, can be a problem for smaller lumber yards. Many smaller yards do not have forklifts capable of lifting the greater weight larger units. In addition, even though a large lumber yard or a distribution yard may have the equipment necessary for the large units, many of their customers may not. Nor may their customers want to tie up money in extra inventory. For these reasons, a discussion of unit size is necessary before an order is written--particularly if a trader knows that the lumber being shipped comes from a mill with large unit sizes.

Units are usually made up of pieces that are all the same thickness and width. Table VI shows the most common number of pieces per unit (called *unit size*) for boards and dimension lumber. Unit size on timbers and factory lumber varies too much to generalize.

TABLE VI

COMMON UNIT SIZES FOR BOARDS AND DIMENSION LUMBER

NOMINAL SIZE	MOST COMMON UNIT SIZE	OTHER COMMON UNIT SIZES
1x4	32x13=416/unit	442, 468, 715
1x6	32x 8=256/unit	272, 288, 440
1x8	32x 6=192/unit	204, 216, 330
1x10	32x 5=160/unit	170, 180, 275
1x12	32x 4=128/unit	136, 144, 220
2x4	21x14=294/unit	208, 260, 273
2x6	21x 9=189/unit	128, 160, 168
2x8	21x 7=147/unit	96, 120, 126
2x10	21x 5=105/unit	80, 100, 105
2x12	21x 4= 84/unit	64, 80, 84

10. How Many Lengths Per Unit?

Not all lumber is produced in the same lengths. Table VII is a listing of normal **lengths** that lumber is produced in, by lumber type.

TABLE VII

NORMAL LENGTHS PRODUCED, BY LUMBER TYPE

LUMBER TYPE	NORMAL LENGTHS PRODUCED
Boards	6' through 16', some mills to 20' (2' increments, even only)
Dimension	8' through 16' or 20', some mills to 24' (2' increments, even only)
Factory	6' through 16' (1' or 2' increments)
Timbers	8' through 40' (1' or 2' increments)

Note: Most grade books say that for each of the four types of lumber listed above, lengths may be sold 6' and longer, and in 1' increments. In practice, however, the lengths and increments in the table reflect what is normally traded.

Although a unit of lumber usually contains pieces of the same thickness and width, it may or may not contain pieces all the same length. The number of different lengths that a mill will unitize together is generally determined by the number of sorting bins that the mill has for stacking the finished lumber. Each bin holds lumber that is all the same grade, thickness, and width. A large, modern sawmill may have as many as thirty or forty bins into which lumber can be electronically sorted. This means that the mill can separate every length in at least the main grades. A smaller sawmill, or a mill with fewer sorting bins, may have to put two or more lengths together in a bin, and therefore, the unit. Some mills have so little space for sorting that they must unitize several lengths together.

The most common groupings of lengths per unit are:

1. pulled to length, which means that all of the pieces in the unit are the same length;
2. two lengths per unit, usually 6" and 8", 10" and 12", 14" and 16", and 18" and 20";
3. 12" and shorter together, 14" and 16" together; and
4. random lengths, all lengths together.*

As a rule, dimension lumber is pulled to length. With the other types of lumber (boards, timbers, and factory) the configuration of lengths per unit must be checked with the mill.

11. How Is the Lumber Protected?

Unless otherwise stated, all lumber being traded is assumed to be *clean and bright* (or *fresh run*). Clean and bright is the appearance condition of lumber as it comes out of the planer at the mill.

To ensure that lumber arrives at its destination clean and bright, and also to ensure that dry lumber does not pick

---

* These units usually include "nested shorts." Nested shorts are the short lengths placed end-to-end to make the equivalent of a longer length (e.g., a 6" and an 8" foot piece butted end-to-end to make a 14" piece.) Nested shorts make the units more square and increase the footage per unit, but are impossible to tally at the receiving end because whether the inside pieces are solid or nested shorts cannot be determined.

up excessive moisture either while traveling to the destination or while being stored at the destination, units of lumber are frequently protected from the elements. Three conventional methods are used for protecting lumber during this time of shipment and storage. They are paper wrapping, poly under the top tier, and chemical protection. A description of each follows.

Paper wrapped. Paper (or sometimes plastic or poly) wrapping is the most common form of protection for units of **dry** lumber. Green lumber is seldom wrapped.

Paper wrapping is the process whereby the top, sides, and ends of a unit are wrapped with a heavy, fiber-reinforced, all-weather paper. This paper is usually opaque and printed a light color to reflect heat and direct sunlight. The best paper also contains tiny holes in the part that covers the sides. These holes allow evaporated moisture to escape. (Without these breathing holes, evaporated moisture collects on the outside pieces of lumber, encouraging mold.)

Being paper wrapped, units of lumber can travel on flatcars or open trucks yet still arrive at the destination clean and bright. Paper wrapping also enables outside storage at the destination without deterioration of quality.

With the exception of the very lowest grades, (i.e., economy and #5 common), most dry lumber that regularly ships on flatcars or flat-deck trucks is paper wrapped. This includes dimension lumber in the grades of standard and better, #2 and better, and usually the grades of utility and #3. It also includes most factory lumber and the upper and middle grades of boards; that is, always #2 and #3 common and usually #4 common.

Because of the many different ways that lumber is transported, the numerous grades and items produced, and the different ways that lumber is stored at destinations, most mills do not automatically paper wrap their units. Most are, however, equipped to do it for an upcharge of between $5 to $8 per MBF. A notable exception to this is Canadian dimension mills, who **do** usually paper wrap most random length dimension, and include this in the mill sales price.

PUTT (Poly Under the Top Tier.) As dry, surfaced units are being assembled in the mill, each complete *tier*, or horizontal row, is stacked one on top of the other. If lumber is to be PUTT protected, just before the final tier of lumber is laid, a piece of clear plastic is stretched over the unit. The top tier is then put on and the steel bands are wrapped around and tightened. In this way, the plastic does not protect the top layer, the ends, or the sides; but it does keep rain from running down through the middle of the unit. PUTT is usually used on dry dimension lumber of lower value (e.g., utility and #3).

Chemical Protection. There are several types of liquid chemicals that can be applied to lumber by dipping or spraying. These treatments can be done to either individual pieces or to entire units of lumber. These chemicals impede the growth of mold and fungus and provide a chemical barrier against the elements. This protection prevents discoloration and the absorption of moisture so that no additional wrapping is necessary. Called *anti-stain* and *anti-travel* protection, these treatments are used almost exclusively in the export market; seldom in the domestic market.

When lumber is not protected in any way, the units are said to be *open*. If lumber is going to travel in a boxcar and be stored under cover upon arrival, no protection is necessary. Studs, high grade boards, and factory lumber frequently ship this way.

Most green lumber is also shipped as open units, even when it is being transported on flatcars. Green lumber has so much moisture to lose that wrapping it would increase the chances of discoloration from stain and mold. Economy and #5 common dry lumber also travel on flatcars without protection because the value of the lumber is low and its appearance is not critical.

WHAT, WHEN, AND HOW IS THE LUMBER TO BE SHIPPED?

12. What Is the Approximate Quantity?

It is important to know early in a lumber-trading conversation, what quantity of lumber the customer wants or can use. Quantities for a given order can range from a few thousand board feet (a partial truckload shipment) to 115,000 board feet (a large flatcar shipment).

The quantity and the type of lumber to be shipped will affect the shipping method. As a general rule, the larger the quantity, the lower the cost of transportation per MBF. Listed in Table VIII are the various conveyances normally used to move the listed quantities of lumber.

TABLE VIII

TRANSPORTATION OPTIONS
FOR VARIOUS QUANTITIES SHIPPED

QUANTITY	TYPE LUMBER	CONVEYANCE
1-22 MBF	Most dry or green lumber	LTL (less than truckload)
22-30 MBF	Most dry lumber	full truckload or pig van
45-60 MBF	Dry Dimension	40 ft boxcar
	Dry Boards/Factory	50 ft boxcar
	Green Dimension	60 ft flatcar
	Green Timbers	60 ft flatcar
65-80 MBF	Dimension	50 ft boxcar
	Dry Dimension	50-60 ft flatcar
	Dry Boards/Factory	56-60 ft flatcar
	Green Dimension	60-73 ft flatcar
	Green Timbers	60-73 ft flatcar
85-115 MBF	Dry Dimension	66-89 ft flatcar
	Dry Boards	66-89 ft flatcar

13. What Is the Tally?

The *tally* is the number of pieces, or the number of units, of each different size, grade, and length of lumber on a particular order. Tallies are a critical part of any lumber trading discussion because both the sawmill and the customer have specific tally needs in mind when they sell or buy. What might be acceptable as a tally to one may not be agreeable to the other.

In general, customers need sizes of lumber based upon the end use of that lumber. Sawmills, on the other hand, need to maximize volume and grade of lumber output, based upon the different sizes and quality of their logs. Sawmills have to sell everything that they produce, not just the high-demand premium items. Because the sizes preferred by the customer frequently do not match those sizes the sawmill has to sell, negotiating a tally acceptable to both the customer and the sawmill is critical.

Let us look at an example. Most 2x4 dimension buyers like to buy 2x4 all 16 foot. This is a length that works well when used in conjunction with other building materials, such as plywood. Sawmills, however, while producing 16' 2x4s, also end up with 14' and shorter 2x4s. A mill needs to sell its entire production of 2x4s, not just the 16 foot. It is the job of the trader, then, to: find out how many 14' and shorter pieces a customer will take in order to get the required number of 16s; find out how few 14' and shorter pieces the mill is willing to sell in order to ship the necessary 16s; and work out an agreeable compromise. Because this is important to successful trading, there is more about tallies in Chapter V and Appendix A.

As a rule, lumber orders include just one **type** of lumber. A discussion over an inquiry of boards with a customer, for example, may include some talk about different widths, and perhaps different lengths, but it will probably not include talk of dimension lumber. Similarly, a discussion about a dimension-lumber order will likely include different lengths, but it will probably not include any mention of factory lumber or timbers.

14. What Is the Loading?

The *loading*, in lumber industry terms, means the type of equipment used to transport the lumber. Since the quantity of lumber shipped directly affects the total weight of the shipment, and because freight bills are usually based on weight, it is necessary to match the quantity (and therefore the weight) being shipped with the appropriate conveyance (see Table VI). The type and the value of the lumber being shipped may also help to determine the loading.

The most commonly used types of equipment for hauling lumber are: flat-deck trucks, piggyback vans, boxcars, flatcars, and all-door railcars. The following is a description of these kinds of equipment, and the type of lumber most often transported in each.

Flat-Deck Trucks (Flat Decks). A flat-deck truck is simply a platform on wheels, pulled by an over-the-road truck tractor. Flat-deck trucks are completely open on the top and sides so that units of lumber can be easily loaded and unloaded with a forklift. Flat decks haul any kind of lumber that is unitized. A fully-loaded flat-deck truck will typically carry 48-50 thousand pounds of cargo, which is approximately 26-30 MBF of dry boards or dimension lumber. Tarps to cover the lumber in transit are generally available for an additional charge (usually a flat charge of $25-$40 for the load).

Piggyback Vans. As described in Chapter 3, pig vans are the approximate size of over-the-road truck trailers, but are open for loading only in the back. This means that the loading of a pig van requires a ramp for forklift access.

Pig vans generally haul studs, short-length items, and items of high value (like cedar siding and high-grade boards and factory lumber). The loading of random length material into a van is more difficult than the loading of a flat-deck truck, but because the rates are frequently cheaper than over-the-road trucking, more and more lumber is being shipped this way. A fully-loaded pig van of boards or dimension lumber carries approximately the same weight and quantity as a flat-deck truck--that is, 26-30 MBF.

Piggyback Flats. Flat pigs, as they're referred to, are similar to pig vans, except they are flat-decks instead of vans, and they are loaded and unloaded like flat-deck trucks. Flat pigs can haul anything that is unitized and, usually, paper wrapped. (Paper wrapped because tarping is seldom available for flat pigs.) Flat pigs are not used regularly because they are not readily available.

Boxcars. Boxcars are fully enclosed railroad cars. They have either single or double sliding doors on each side for loading and unloading. Single-door boxcars generally have a 10' to 14' opening; double-door boxcars have a 12' to 16' opening. Because they provide full protection from the weather, boxcars are used for transporting lumber that is not paper wrapped and will be stored under cover at the destination. This includes random length high-grade boards and factory lumber, studs, and short-length items. Loadings in boxcars are frequently referred to as UDDB - meaning unitized in a double door boxcar. Boxcars are also used for carrying *loose loaded* (not unitized) items. A fully-loaded 50' boxcar of studs will typically hold 65-70 MBF. When filled with R/L board items packaged two lengths per unit, it will generally haul 48-52 MBF.

Flatcars. Flatcars are railcars fully open on the top and sides. Because they offer no protection, most dry lumber shipped on flatcars is paper wrapped. Flatcars used for lumber range in length from 48' to 89'. Flatcars are commonly used for shipping timbers, random length dimension lumber, and middle- to low-grade boards and factory lumber. The loading onto and off these cars can be done from one or both sides.

There are many varieties of flatcars. In addition to regular flatcars, two special types of flatcars that are frequently used for hauling lumber are: *A-frame* (also called *center-beam*) flats, and *bulkhead flats*.

A-frames, or *center-beams*, are flatcars that have a solid divider running down the middle, the length of the car. This divider is slightly higher than the lumber when the car is loaded. Because the lumber can be secured to the center beam as well as the deck of the car, less bracing and strapping is required, yet the load is very stable in transport. Centerbeams are faster to load and unload, and, because they require less strapping, they are less expensive to load. However, the solid center divider necessitates loading and unloading from both sides of the car. While this is not a great problem for most sawmills, it can be a problem for customers, many of whom have yards that permit access to only one side of a car. For this reason, before lumber is loaded on A-frame (center-beam) cars, customers must be asked if they have the facilities for unloading from both sides. A fully loaded 66' center-beam may carry as much as

110 MBF of one-length-per-unit dry dimension lumber, or 90-95 MBF of two-lengths-per-unit boards.

Bulkhead Flatcars (BHFC's) are like normal flatcars except that they have a vertical flat wall at each end. Called bulkheads, these walls keep material from shifting front to back as the car starts and stops along the way. As these walls cause no additional problems for loading or unloading, bulkhead flatcars are often preferred by customers for the slight added protection they offer. A 56' bulkhead flatcar filled with two-length-per-unit-boards will hold approximately 65-70 MBF. Filled with dimension lumber, it will carry 80-85 MBF.

All-Door railcars. These cars are a hybrid between a boxcar and a flatcar. All-door cars are like boxcars except that their side walls are actually sliding panels. Like closet doors that can be shifted all to one side or the other, all-door cars are loaded and unloaded with the ease of a flatcar. Once an all-door car is loaded and ready to travel, the panels slide closed and the lumber travels in the protection of a boxcar. All-door cars are usually owned or leased by individual sawmills that use them to help promote sales. Because of their obvious advantages, these cars are in high demand and sometimes carry an additional charge for their use. All-door cars carry anything that is unitized. An all-door car can hold approximately 55-58 MBF of 8' studs, and approximately 53-56 MBF of boards.

15. What Is the Shipment Time?

Every customer wants to know when to expect lumber that has been ordered, and so the shipment time must be discussed. *Shipment time* always refers to mill shipment time. That is, it refers to the time when the lumber will leave the mill. Standard sawmill shipment time for a back-to-back order is anywhere from "10 days or sooner", to "30-45 days."

When an order is being discussed, the mill will indicate the shipment time, usually to the nearest week, based upon its production and shipment schedule. Because in most cases customers want their lumber as soon as possible, shipping instructions are usually written on an *or sooner* basis, meaning that if the lumber is available sooner than the stated week, the mill will to ship it. Occasionally a customer will request that lumber not ship earlier than the stated week, in which case the order must include the instructions *no sooner.*

Because mills frequently sell projections of their output, and not simply lumber already produced and awaiting shipment, it is not unusual for some orders to ship late. Deviations in scheduled shipment time up to four or five days beyond the agreed week are generally accepted in the trade, even though, technically, such an order is out of contract and a shipment claim could be filed.

A trader needs to keep track of the timing of all orders. Once an order is late in shipping, the trader needs to find out from the mill when it will ship, and tell the customer. Once an order has shipped, the car number or van number and the shipment date should be communicated immediately from the mill to the trader to the customer.

On average, trans-continental railroad transit time is two to three weeks. Trans-continental pig-van transit time is slightly faster. Trans-continental truck transit time is five to seven days.

Shipment time is discussed in terms of sawmill shipment, not arrival at destination, time because once a lumber order is loaded and consigned to the carrier- whether rail, pig-van, or truck- neither the trader nor the sawmill can control its speed. For this reason, a trader should never promise any particular arrival at destination time.

Orders in transit can be slowed for many reasons, and carriers do not generally notify the trading company when this happens. In the case of a rail or van shipment, a particular car may be stalled in a railroad switching yard, it may be slowed by bad weather, or it might be *bad-ordered* (meaning that the car is broken or unable to proceed until repaired). Alternatively, a car may go straight through to the destination and arrive sooner than expected.

In the case of a truck shipment, slowdowns can occur when a truck encounters mechanical problems, when weather is bad, or when a driver is unreliable.

If there is particular urgency on a specific shipment, the trader can periodically *trace* a car, truck, or van by calling the carrier and finding out its current status and location.

16. What Is the Customer's Delivering Carrier or Truck Delivery Address?

Because a lumber price is quoted as a delivered price including the freight, and because different railroads have different rates, it is necessary to know a customer's delivering carrier before calculating and quoting a delivered rail price. If a particular customer is on the Union Pacific Railroad, for example, it may be cheaper to quote lumber from a sawmill also on the Union Pacific, even if the mill price of the lumber is higher than comparable lumber from a mill served by another railroad.

To ensure that the best possible price is quoted, it is necessary to know whether a customer taking rail shipments is *open to reciprocal switch*, (or simply *open*). Reciprocal switching is an agreement worked out between a customer, located on one railroad, and the other rail carriers serving that customer's town. If a customer is open to reciprocal switching, railcars can be delivered (or switched) to the customers yard from another railroad without paying an additional *switching fee*.

For example, let's say that a customer is located on a Burlington Northern (BN) siding in a town served by both the Burlington Northern and the Union Pacific (UP) railroads. The sawmill you wish to ship from is located on the UP. After checking the various rates, you find that the UP direct rate is the cheapest rate available to that destination. If the customer is open to reciprocal switching, the lumber can be shipped into the customer's town on the UP and be delivered to the BN siding without additional charge.

Not all customers will be able to negotiate a reciprocal switching agreement with the other rail carriers serving their location. In the example above, if the customer were not open to reciprocal switching, lumber must be either shipped to that town on a higher rate that includes the BN in the routing, or shipped on the UP, but with the customer paying an additional BN local switching fee. (A switching fee may amount to as much as several hundred dollars.)

In the case of truck or pig-van shipments, reciprocal switching is not an issue. It is necessary, however, to have the correct street address for deliveries. This is sometimes different from a customer's mailing address. A truck or van sent to the wrong location will increase delivery costs unnecessarily.

THE PRICING AND TERMS OF PAYMENT

17. What Is the Price?

Lumber is normally quoted on a delivered price basis. Therefore the quotation price will include: the sawmill price of the lumber plus freight plus profit. All this is expressed in terms of a price per MBF. Any other additional charges, such as tarp charges, special handling or remanufacturing charges, import or export fees, and any provincial or federal sales taxes (in Canada only), must be included and stated clearly, or they must be itemized with the understanding that they are in addition to the quoted price. Most lumber traded in North America is transacted in US funds, unless the destination is Canada. In that case, the business is usually done in Canadian dollars.

18. What Is the Customer's Credit Status?

Every customer has a credit limit. It is the responsibility of the trader to verify, before a sale is made, that the customer's credit has been checked and approved. There is usually someone in each trading company who is responsible for establishing and maintaining credit records. A customer's payment history, together with information from outside credit agencies, can help to determine how much credit should be extended to each account.

If a customer with a good payment record wishes to make a purchase above his credit limit, it is sometimes possible to take this additional business based upon conditional credit approval. This conditional credit approval is usually done in a case where it can be shown that earlier orders will have been paid by the time that the current order arrives or is ready to ship. Conditionally approved orders going by rail or van may be shipped to the customer's destination, but are always kept in the shipper's (the trader's company's) name until the expected payments are received and full credit approval is given. In this way, if the customer should go out of business, or for some other reason should become credit unworthy, the trader, as the shipper, has the option to stop the car (or van) and resell it somewhere else.

19. What Are the Terms of Payment?

Terms of payment in the industry vary widely depending on the type of account. Wholesale accounts generally have terms based upon the *date of invoicing*. Retail accounts usually have terms based upon the arrival date of the material. Whether wholesale or retail, a cash discount is usually given to encourage prompt payment.

As a rule, invoice terms contain the following elements:

1. The percentage of discount allowed for prompt (usually called "cash") payment. This element is referred to as "% CD" (the percent cash discount). The most common cash discount is 2 %.
2. Deduction of the cost of freight from the gross amount invoiced, referred to as "ADF" (after

deducting freight). This deduction is necessary because rail and truck carriers do not give a cash discount for prompt payment; therefore, no discount on this portion of the invoiced amount can be allowed to the customer.

3. Specification of the "prompt payment" element, expressed as a number of days (e.g., 5, 10, or 30 days) or as a particular day of the month (e.g., the 10th of the month).
4. The date on which the counting of the allowable days for the cash discount begins. This is usually done based upon either the date of the invoice or the arrival of the material. If terms are based upon the date of the invoice, they are called ADI or "after date of invoice" terms. If terms are based upon the arrival of the material, they are called AAC or AAT "after arrival of car" or "after arrival of truck" terms.
5. The option to forego the discount, referred to as the "net" payment terms.
6. The number of days allowed to send the net payment.

Here are some examples illustrating the more commonly-used invoicing terms in the industry:

i. 2% CD, ADF, 10 days ADI, net 15 days ADI;
ii. 2% CD, ADF, 5 days AAC, net 15 days AAC;
iii. 2% CD, ADF, 10th of month AAC, net 15th of month AAC.

With each of these types of payment terms, both the number of days allowed to take the discount, and the number of days allowed for the net due date, are variable and determined by the individual company doing the selling. Because there are differences between different company's terms, and because the terms of payment definitely affect the final cost of doing business, a trader needs to be certain that a mill or customer understands the terms of payment before lumber is bought or sold.

One other set of terms that is starting to be used in the industry is the allowance for a smaller percent cash discount, but the percent being allowed for the **entire** delivered price--including the freight. The reason for this type of terms is both to simplify the invoicing procedure and because of the growing number of different freight contracts and confidential freight agreements being worked out by individual companies. By allowing a discount off the entire delivered price, the seller does not need to show the freight rate on the invoice (and therefore can maintain the confidentiality of the contract rates.) The customer being invoiced, however, receives approximately the same dollar discount as if the freight were shown and deducted in the more normal fashion. Typically, these new terms read:

i. 1 1/2 % CD 10 days ADI, net 15 days ADI;
ii. 1 1/2 % CD 5 days AAC, net 15 days AAC

20. What Is the Customer's Order Number?

Lumber trading is a business conducted primarily on the basis of verbal agreements. An order given, and an order taken, in the course of a telephone conversation, are agreements that are morally and legally binding on both parties. To renege on an order (to back out or disavow an order) is something to be done only at the cost of losing a supplier or a customer. Win, lose, or draw, an order agreed to **is** an order.

The confirmation of an order given and accepted, is its designation with an order number. It is standard industry practice that separate order numbers are used for each carload or truckload being traded. Once order numbers are exchanged, the business is sealed. For this reason one must be very sure that before an order number is given or accepted, one is comfortable with all the terms and conditions that have been verbally discussed or implied. This being done, and the exchange of order numbers being made, the trader will record the mills/customers order number, together with his/her own order number, on the purchase order and/or order confirmation. Any claims, freight bills, order changes, or other communication dealing with this one transaction will show that order number as a reference.

CHAPTER FIVE

Creating an Order

We now know what lumber is, what trading is, how the transportation system works, and all the questions that need to have been considered before buying or selling softwood lumber. It is time for us to put this information to use.

There are scores of books written on how to be a sales person. It is not the purpose of this chapter to recover that ground. There are, however, several issues unique to the lumber industry that should be addressed.

Lumber orders do not just happen. Customers do not normally call up and request a repeat of the last order at whatever the new price is. It simply doesn't happen that way. Most lumber orders are created. They are made to happen by actively soliciting customers and sawmills, and by looking for opportunities to provide a service, profitably.

One of the surest ways to create an order is by taking a position, either long or short, in the market. This means buying lumber long or selling lumber short. By taking the correct position in a moving market, a trader can offer lumber for sale at a more competitive price than the competition.

Another way to create orders is to know the markets and take advantage of opportunities for back-to-back orders. An opportunity like this may start with a sawmill that wants to sell some lumber, or it may start with a customer who needs to buy some lumber. In either case, a trader must grasp the opportunity and try to find the offsetting sell or buy to complete it. This matching procedure requires analyzing all of the variables that go into an order - these include the tally, price, species, shipment time, wrapping, etc.- then choosing the best sawmills (or customers) to supply (or to purchase) the material. The trader with the best knowledge of what sawmills can and will produce, the best command of freight rates, and the best idea of what a customer is really going to buy, will most frequently end up with the order.

To most successfully capitalize on these opportunities, a trader needs to ask lots of questions and to discuss different options with both the customer and the sawmill.

These discussions frequently boil down to trading off price versus tally. It is around these two key issues, price and tally, that a resourceful trader can create many orders. The following is a discussion of these two related issues.

PRICING

Sawmills price and sell their lumber independent of each other. Yet, because of the relatively free flow of information between buyers and sellers throughout the industry, approximately equal industry-wide prices exist for each width, and, in the case of dimension lumber, even each length of each width. If a certain mill happens to be cheaper than its competition on a given item, it will quickly sell out, or it will raise its prices. If that mill is more expensive than its competition, it will sell little or adjust its prices downward. Since the price for every item is determined solely by supply and demand, these prices are constantly shifting up and down.

Besides the daily interchange between buyers and sellers, estimates of market prices can be determined by checking published pricing guides. Because prices are so volatile, and because both buyers and sellers need to keep track of items they are not necessarily trading every day, several price-reporting newsletters are published weekly or biweekly. The writers for these newsletters survey sawmills, wholesalers, distribution yards, and retailers throughout the industry. They report their price findings by species, grade, and size for the most widely-traded board, dimension, timber, and factory items. For the main dimension lumber species and grades, prices are even reported by specific lengths. These publications are an important tool to the industry. Mills, buyers, and traders all watch these published price reports to remain current with market levels.

TALLY

Most lumber shipments go directly from one mill to one destination. Picking up lumber at two or more locations, or delivering lumber to two or more locations, involves additional handling and usually considerably higher freight charges.

Most sawmills specialize in just one or two types of lumber manufacturing (e.g. board mills produce boards most of the time, dimension mills produce dimension lumber most of the time, etc.) Of the four types of lumber produced, the great majority of mills produce either boards or dimension lumber as their primary product. For these reasons, lumber orders usually include just one type of lumber. That is, a discussion of tally is usually specific to boards, or dimension, or factory, or timbers.

Each sawmill attempts to produce the maximum number of the most valuable lengths and widths that it can, on each production run. The type and quality of the logs, and the relative market values of the different items on the current market, determine the tally that a mill is able to offer for sale.

Because of this, there is no one universal length tally for any loading, for any size of lumber. This means that when a sawmill offers to sell a R/L 2x10, for example, all that can be assumed for sure about the tally, from this description, is that it will probably include at least one unit of each of the standard lengths (which for 2x10 normally means 8 foot through 20 foot). Every mill will have its own configuration of precisely how many units of each length will be shipped.

Unless otherwise clearly stated, a mill will usually load all the lumber it can on each order. Because mills cannot always control the size of equipment that the railroad, truck, or van carrier will send them for the loading of a particular order, tallies bought from a mill usually include some variations of lengths. This allows for the different configurations of lengths necessary to make maximum loadings on the different equipment sizes.

Understanding all this, we can discuss tallies and pricing, by lumber type, realizing that each type of lumber produced has its own distinct tally options.

Dimension Tallies

Because dimension mills tend to be the largest in terms of production, because most dimension sawmills sort their lumber by width and by length (i.e., packaging only one width and one length per unit), and because dimension mills- particularly those that dry their lumber- usually paper wrap their units, dimension sawmills frequently have sizeable inventories of unsold, ready-to-ship lumber from which to draw special length and width tallies.

Each individual length of each width of dimension lumber has its own particular dollar value. By knowing the values for the different widths and lengths, and having some idea of a mill's available inventory, a trader, or the sales manager for a sawmill, can calculate the price of a very specific dimension tally by evaluating the sum of the lengths involved.

Let's look at an example. Suppose that you have a customer on the phone with whom you are discussing R/L 2x4 at a mill price of $150/MBF. The tally that one mill is offering at this price is: 2/8', 3/10', 6/12', 7/14', 8/16', 2/18', 1/20' (read as: two units of 8', three units of 10', etc.).

The customer's initial response to this tally is that there are too many 12' units and not enough 10' units. You know, by talking to the mills and reading the pricing publications, that the most valuable lengths in 2x4 are, first, 16', then 10'. The lowest valued length is 14', with 12' the next lowest.

You realize, then, that this customer is asking for a tally with fewer low-value items and more high-value ones. Therefore, you know that the customers requested tally is going to cost more than the tally being offered by the mill. By knowing relative length values, you can give the customer several different price and tally options to consider. Substituting 10' units for 12' units, as the customer requested, is certainly one option, but the price will be higher. As you are talking, you can estimate for the customer what the switch from 12' units to 10' units will do to the price. (See the Appendix for calculating the precise cost difference for those additional units.)

In addition, however, you can offer other options. One option, which would allow you to keep the price down, would be to reduce the number of 16' units while increasing the number of 10' units. This reduction in 16s will offset the additional cost of the increased 10s. Another option is to increase the number of 14 foot units while decreasing the 12s. Since 14s are cheap, the cutback on the 12s will be balanced by the increase in the 14s, again leaving the price approximately unchanged.

In doing this, you have given the customer several choices, each of which addresses the main concern for more 10' units, fewer 12' units, and the implicit concern for a low price. As you discuss the various choices, you get a much better sense of what the customer really will buy and at what price.

After the tally has been agreed to and an order has been concluded, the tally part of the purchase order to the mill and the order confirmation to the customer might look something like this:

> 1 C/L (carload), 72-76 MBF of 2x4, R/L, 8'-20', to contain: 2/8', minimum 4-5/10', maximum 4-5/12', 8/14', 7-8/16', 2/18', 1/20'

Boards Tallies

Board mills are generally smaller than dimension mills. This means that their production runs and the volumes produced are smaller. Because board mills do not generally wrap their lumber, they can't store it outside for any length of time, and therefore do not build large inventories.

As a rule, board mills do not unitize in single lengths (pull to length). Instead, they package two, three, or four lengths per unit. These multiple-length units make it much more difficult to be specific on a length tally. Even the mill sales manager won't know the composition of individual units without physically going out and reading the individual *unit tally* off of each unit. (A unit tally, stating the exact number of pieces of each width and length in the unit, is written on a small piece of cardboard and stapled to the end of each unit. It is from these unit tallies, recorded by the loading foreman as the unit is loaded on the railcar or truck, that the total tally for the shipment is generated). Therefore, length tallies purchased from board mills are generally only estimates.

There is a general pattern, however, to what can be expected from a board mill's length tally: if a board is good enough to be used 12" wide, it is likely also to be good enough to be used 14' or 16' long. If a board is only good enough to be used 4" wide, there is a good chance that it will be fairly short. Following this reasoning, board tallies tend to have more longer lengths in the wide widths, and more shorter lengths in the narrow widths. The middle widths produce the most equal cross-section of lengths.

Due to these factors then: smaller volumes produced, stock not pulled to length, and no sizeable inventories from which to draw, width and length tallies from board mills are not as specific as those from dimension mills.

The tally part of a purchase order or customer confirmation for a board order might look something like this:

1 C/L (carload), approximately 65-70 MBF, to contain:

12-15 MBF	1x4	6'-16'	good assortment of lengths
12-15 MBF	1x6	6'-16'	good assortment of lengths
15-18 MBF	1x8	6'-16'	good assortment of lengths
10-13 MBF	1x10	6'-16'	good assortment of lengths
7-10 MBF	1x12	6'-16'	good assortment of lengths

As you can see, the mill is given a much greater degree of flexibility in the loading of this carload of boards than for a mill loading a carload of dimension lumber. With board mills, just as with dimension mills, this flexibility is also partly necessary because of uncertainty of transportation equipment size.

Factory Lumber Tallies

Tallies on factory-lumber orders are even less specific than those for boards. Because production runs are even smaller than board runs, the tallies on orders for factory lumber are sometimes not even specified to grade--much less to width and length. Factory lumber is usually priced by grade, and approximations of the percentage breakdown by grade is given, but it is only when the lumber comes out of the planer, and is graded and tallied, that the actual grade breakdown is known.

For example, an order of shop lumber might read as follows:

1 T/L (truckload), approximately 28 MBF, to contain:
4/4, #2 & Btr Shop, S2S 7/8", H&M, RW/RL
#2 Shop @ $285/MBF
#1 Shop @ $380/MBF
#3 Clear @ $515/MBF

In plain English, this translates as follows:

4/4 - reads as "four quarter" and represents the nominal thickness or *count* on which the lumber will be invoiced.

#2 & Btr Shop - means that the grade will include: #2 Shop, #1 Shop, and #3 Clear. Each piece of lumber will be graded and marked. A specific price is assigned to each of the three grades.

S2S - means that the lumber is only surfaced on the two wide sides

7/8 - means that the net thickness will be 7/8" instead of the standard 3/4"

H&M - stands for hit and miss and means that because the lumber is being surfaced thicker than normal,some areas on the face will not be surfaced clean and smooth (they will be missed by the planer).

RW/RL - means random width, random length. Normally widths will run from 4" to 12" (sometimes wider), and will typically be the heaviest (the most) in 6" and 8". Lengths will usually run from 6' to 16', light (the least) on the 6' and 8' and heaviest on the 12' to 16'.

Because factory lumber is bought to be remanufactured--that is, cut into smaller pieces--many factory lumber orders can be flexible like this. The value for the different grades is based upon what can be yielded from each grade. In this way, the price reflects the yield. The lower grade lumber is cheaper but yields less, the higher grade is more expensive but yields more.

Timber Tallies

Tallies on timbers are generally more specific than those for boards, but not as specific as those for dimension lumber. Lengths on timber tallies generally start at 8' and can go up to 40' or more, depending on the setup of the mill. Timber tallies are usually pulled to length unless the quantities are small, in which case multiple lengths may be together in a unit.

The demand for very large sizes of timbers is so unpredictable that many timber (sometimes called *cutting*) orders are not produced until the sawmill receives the order. Occasionally specific trees are actually felled for specific large timber orders.

KEY PHRASES

In a discussion with a sawmill or a customer about a potential order, a trader must constantly be asking questions to obtain enough information to properly price and get the order. There are two phrases that are used widely throughout the lumber industry and require some explanation. These phrases are "firm offer", and "subject to prior sale."

Firm offer. The term firm offer has tremendous meaning in the lexicon of lumber trading. Simply put, a firm offer is a *counter* offer. It is promise to sell or buy the lumber in question, if certain conditions are met. The conditions to be met may be anything having to do with the lumber, from shipment time or tally, to the price or loading.

A firm offer works like this: a trader and a customer are discussing a particular carload of lumber that will deliver to the customer's yard for $255/MBF. The customer agrees with everything about the carload except the price.

The trader might then ask, "If I can get this lumber to you at $252/MBF, will you buy it?"

"I really was hoping to buy this lumber at $250/MBF," the customer responds.

"OK," says the trader, "Will you give that to me as a firm offer at $250/MBF? I'll go back to the mill and see if they'll take a slightly lower price."

"Yes, you can have that as a firm offer at $250/MBF until 2:00 this afternoon," the customer agrees.

In getting this firm offer, the trader has a promise from the customer to buy the carload, just as discussed, at

$250/MBF. Knowing this, the trader realizes that if a lower price can be negotiated with the sawmill, the buy order between the mill and the trader can be completed. It then only remains to go back to the customer to formalize the sell order.

A firm offer is so important because while a sawmill will have an asking price for each item on its list, that price may not be the cheapest price it is really willing to sell at. The mill will not offer lower than its asking price without a bona fide order for fear of driving the price even lower. In other words, to buy at a cheaper price than the mills normal asking price, the trader must be prepared to actually place the order, should the mill decide to accept the lower price offer.

In this example, a firm offer is given to obtain a particular lower price. Firm offers, however, can be used for any reason. Maybe a customer wants an order to ship within one week, and will give a firm offer on that basis, or perhaps a customer wants the lumber in an all-door car, and will give a firm offer on that basis. Firm offers can be given on tally, on quality of stock, on type of loading, on anything that affects the lumber or the way that it is shipped.

A firm offer is the same as an order, if the specified conditions are met. To renege on a firm offer is like reneging on an order; it simply is not done.

Because of the negotiating leverage that it gives with a sawmill, a firm offer is an important tool that should be used wherever and whenever possible.

Subject to prior sale, or *sub prior*. Like firm offer, the phrase subject to prior sale (or simply sub prior) has great significance in the lumber business. It is used to protect the trader from things beyond his/her control.

Because of the unique characteristics of every different lumber order, until it is actually finalized, each discussion about an order must contain, either explicitly or implicitly, the idea that the lumber is being offered on the condition that it is, in fact, still available for sale. Between the time it takes to discuss an inquiry with a sawmill, call the customer, sell the order, and then call back the sawmill to place the order, the mill may have sold that very lumber to someone else. A trader cannot be on the phone to both the customer and the mill simultaneously. Therefore, unless the mill has agreed to hold the lumber for some period of time, no guarantees can be made that the lumber in question is still available.

Here is an example of how the phrase "sub prior" might be used: A trader is on the phone to a customer. The customer requests a price quotation on some particular type and tally of lumber. The trader writes down all the customer's requirements, asking questions and getting just as many options as possible. The trader hangs up the phone then calls the sawmills that may be able to quote a price for this inquiry. The trader may need to call five, six, or more different mills to check for the best *coverage* (the price, shipment time, tally, etc., that a mill will take an order at). The trader, when talking to the different mills, comes as close as possible to the customer's requested specifications. At the same time, the trader gets other additional options that may be of interest to the customer. After making all the mill calls, the selling trader compiles the information and selects the one or two options that come the closest to what the customer has requested.

The trader calls the customer back and offers the best of these options. It is at this point that the trader must relate to the customer that the material is being offered sub prior, or subject to prior sale. The trader does not yet own the lumber, and can only offer it for sale on the condition that the sawmill has not sold the lumber to someone else. If the customer is interested and wants to buy the material, the trader gets a firm offer, then goes back to the mill and tries to buy the lumber.

To people who have been in the industry for some time, the assumption is usually made that all coverage is offered subject to prior sale. However, it is a good idea to reiterate this phrase, especially when talking with a new customer or trader.

By listening carefully to what a customer really wants to buy (or a sawmill really wants to sell), by being creative and open minded about how best to arrange the order working with price and tally, and by getting a commitment from the customer in the form of a firm offer, a creative trader can land profitable orders while at the same time providing service to both customers and sawmills.

CHAPTER SIX

Handling Claims

A lumber claim arises when the lumber received is not, in the customer's estimation, the lumber that was ordered. If, for any justifiable reason, a customer feels that the lumber arriving is different from what was specified in the original agreement, the customer can file a claim for compensation. This claim can result from something as major as the lumber being *off grade* (not conforming to grade), or as minor as the wrong color of end paint.

The most common areas in which claims arise are: transportation, general grade or manufacturing, moisture content, tally and shortage discrepancies, shipment time, and market movement.

A discussion of each of these types of claims will follow later in the chapter.

If a trader's buy order to the sawmill and sell order to the customer are identical (as they should be), then a claim is really a dispute between the customer and the sawmill (or in some cases, the transportation company). When a claim arises, it is the trader's job to act as the mediator. Since only the trader has access to both disputants, it is up to the trader to keep the parties negotiating until a solution is found that is acceptable to all. Improper handling of a claim can cause the trader to lose credibility with the sawmill, lose the confidence of the customer, and lose money when arbitrary deductions are taken by the customer from the invoice, not to mention the anxiety and wasted time that invariably accompany a bad claim.

WHAT TO DO WHEN YOU GET A CLAIM

Most claims arise out of a real, or at least what is perceived as a real, problem with lumber after it has arrived at a customer's yard. Claims don't go away. At the first indication from the customer that there is a problem, you, as the trader, must realize that **you** have a problem. You must immediately define the problem and get the customer to document it. You must also talk with the customer about what compensation will help resolve the problem. You need to present a clear and realistic description of the situation to the sawmill. You are the customer's representative to the mill.

Likewise, you must represent the position of the sawmill to the customer. The mill may be able to produce documentation verifying its position on the issue. It is your job, then, to transmit this information to the customer. Finally, through the process of negotiation and arbitration, you must conclude a settlement acceptable to all parties (and acceptable to you).

Situations will arise in which the mill says one thing, the customer says another, and you are stuck in the middle. Frequently in a case like this, you (that is, your company) will already have paid the mill's invoice, yet the customer has not paid the invoice to you. You and your company's money is outstanding, and it is your skill as a negotiator that will determine how soon and how much money is recovered.

WHAT ARE YOUR OPTIONS?

There are a number of possible outcomes to every lumber claim, ranging from most desirable to least desirable. Most desirable is the customer agreeing to accept the material and paying the invoice in full. Equally desirable is the sawmill accepting the problem as its own fault and offering an acceptable settlement. The middle ground is compromise, when neither the customer nor the mill can prove conclusively that the other is at fault. This compromise settlement usually takes the form of a slight price reduction by the mill, with the customer agreeing to keep the lumber as it is.

If a compromise is not possible, a *reinspection* can be called for. A reinspection means that a representative of the mill's grading association goes to the customer's yard and actually, piece-by-piece, re-grades the entire shipment of lumber to determine whether or not it is up to the grade and the conditions of the order (more on this later). Finally, and least desirable, is when the customer refuses the lumber, and it must be removed and resold to another yard.

These last two options, a reinspection or a resale of lumber after it has been refused, are last-course alternatives. If the buy order to the mill and the sell order to the customer are in complete agreement, these alternatives are rarely necessary.

Regardless of the seriousness of the situation, however, if there is a claim problem with lumber upon its arrival, it is the responsibility of the customer, as part of the buy/sell agreement, to unload the material. The customer must keep the lumber intact and, preferably, under cover while the normal grievance procedure is immediately begun.

HOW SHOULD YOU PROCEED?

There are two cardinal rules to always keep in mind when working through a claim, regardless of the particulars of the claim. **Rule number one is: Settle the claim as expeditiously as possible.** You must keep on top of a claim. Make sure something is always happening to move that claim toward the final settlement. Something happening may mean that pictures are in the mail; that the mill is discussing the problem with its production people; or that the customer is going out to look at the lumber one more time; but you, as the middle party, must ensure that progress is being made. Claims only get worse with age.

Rule number two is: Document. The customer must substantiate grievances with physical evidence. This evidence may be photographs--good photographs that will be self-explanatory to the mill; or it may be samples of the problem lumber; it may be a notarized tally (for a tally claim); or it may be a railroad inspector's report (for a transportation claim). The required evidence will vary with the type of claim. Whatever the claim, however, the customer must prove the case with evidence if monetary compensation is being sought.

Having the physical evidence and documentation sent right away is very important. Even under ideal conditions, pictures, samples, etc. will take a week to ten days to get from the customer to you and on to the mill. With each passing day, a satisfactory settlement becomes more difficult.

A customer filing a claim must also clearly define for you the loss. If a claim problem has cost the customer an additional number of work hours, document this. If lumber has been broken or damaged, document this. If there are other additional costs incurred by the customer, they should be documented. This documentation does not necessarily mean that all costs are recoverable, but if they are not documented, they surely will not be recovered.

The final documentation is your own. You should start a claim file the minute a customer initiates a claim. This file will document progress and the current status of the claim. This file should include notes from conversations with the customer and the mill, as well as copies of letters or evidence that has passed between the customer and the mill.

Letters to both the customer and the mill should go out just as soon as a claim is received. These letters put everyone on record as to the existence of the claim. They describe your understanding of the problem, and they project your outline of the steps to be taken to settle the claim.

TYPES OF CLAIMS AND HOW TO HANDLE THEM

Keeping rules number one and two firmly in mind, what you do next with a claim depends upon the type of claim. The following is a description of the types of claims outlined earlier, together with an explanation of how and why each of these claims can happen, and some suggestions for getting each resolved.

A Transportation Claim

If a claim is a problem that is caused by the transportation carrier or its equipment while the lumber was en route from the sawmill to the destination, the transport company should be notified immediately, **before the material is unloaded.** In a claim such as this, you will be negotiating with the carrier and attempting to win compensation from them. This is the one time that you must be able to show proof of the problem before the lumber is unloaded.

The most common problems with railroad equipment are caused by either improper loading or bracing by the mill, or *humping* (railroad cars being banged together, thereby shifting the contents inside). Either of these situations can break the bands that hold the units together so that the lumber is broken, scattered, and mixed. These shifts in the placement of units are called a *load shift*. Load shifts make forklift unloading very difficult, and frequently require additional manhours for unloading. A claim can be made on this basis.

Other carrier-related claims can be caused by, just to name a few: boxcars that are not weather tight; lumber, usually on flatcars, that has been physically tampered with or stolen; railcars improperly placed on the customer's siding; or trucks arriving without tarp protection, so that the lumber has picked up moisture and road dirt.

How to handle a transportation claim. As noted, a problem involving the carrier must be documented while the lumber is still on the car or truck. In a case involving the railroads, rail companies have agents who will inspect a problem car to see if damage was actually caused along the way, or if the problem was the result of improper handling or insufficient blocking and bracing by the mill at the time of loading. Rail companies will generally not accept damage claims unless their representative has inspected the problem while the lumber is still in the car. Even with a railroad inspector on hand, the customer should take accurate and descriptive photographs. With these pictures, you, for the customer, can argue your case if the railroad agent decides that the problem lies with the mill and the way the car was loaded.

On a truck or van shipment, the customer must always sign a bill of lading to verify receipt of the goods. If there appears to be a transportation-related problem with the material, the customer must write a clear and full description of the problem on the bill of lading while the driver is still there to verify, concur, and sign, an acknowledgement of the discrepancy.

A General Grade or Manufacturing Claim

This type of claim arises when, in the customer's opinion, lumber has been misgraded or mismanufactured. Misgraded lumber occasionally happens, because all lumber (with the exception of machine stress rated lumber) is visually graded. That is, it is visually assessed a grade by a person (a grader) as it passes along on a moving chain. This visual grading of lumber cannot be considered an exact science. It depends on the judgement of the individual grader at the moment of grading. For this reason, a grade reinspection for a load of lumber allows for a 5% difference of opinion between the regrading inspector and the original grader. This means that in the event of a grade reinspection, more than 5% of the lumber being regraded must be off grade before the mill is considered to have lost the reinspection.

Mismanufacturing happens when machinery at the mill is not adjusted or sharpened properly and the finished lumber is not sized or surfaced as it should be. Mismanufacturing can result in lumber that is too thick or too thin, too wide or too narrow, lumber that is not uniform in size or surfacing, or lumber that contains gouges or has wavy (uneven) surfacing.

When a claim appears to be one due to off-grade or mismanufactured lumber, the customer must file a complaint within ten days of receipt of goods. (If the lumber is wrapped or unitized the time limit is ninety days, but only on those characteristics not altered by time, such as knot structure, wane, uneven surfacing, etc..)

How to handle a grade or manufacturing claim. When handling a grade or manufacturing claim, first (after being sure that the stock is unloaded and stored in a protected area) make certain that the customer understands exactly what characteristics are allowable in the grade. Many customers who buy lumber regularly do not know precisely what is allowable in the grades they buy.

Next, get an idea of what percentage of the material the customer feels is below grade, and discuss with the customer whether he has a market for this lower-grade material. Then discuss the problem with the sawmill. The sales person for the mill may have some ideas as to why the grade could be off. If this is the case, the mill may be willing to negotiate a reduced price. At this point the mill may request that some pictures or samples be sent.

If, after pictures and samples have been sent, the mill is still sure that the material shipped is on grade, and the customer is still feels sure that it is not, an independent third party can be helpful. This third party, usually someone of the mill's or the trader's choosing (perhaps a different customer in the same area) may go look at the lumber and give an opinion about the grade of the material. The opinion rendered as a result of this casual arrangement is not binding on either the mill or the customer, but it can sometimes help convince one party or the other that the lumber is or is not on grade.

If both parties still feel the certainty of their position, a reinspection can be called. What is a reinspection? As briefly mentioned earlier, a reinspection means that a representative of the mill's grading association goes to the customer's yard and physically re-grades every piece of lumber in a shipment.

A reinspection may be ordered by any of the parties involved in the dispute; that means the mill, the customer, or the trader. All parties agree beforehand, in writing, to abide by the outcome of the regrade.

After regrading, if it is found that 95% or more of the total volume shipped is of the proper grade, the total shipment is considered on grade and the customer pays for the reinspection (usually amounting to the regrading inspector's time and travel expense). The customer also pays the invoice in full.

In a case where more than 5% of the volume shipped is off grade, the mill pays for the cost of the reinspection. Additionally, all the lumber that is off grade becomes the property of the mill and is deducted from the customer's invoice. The customer, however, must accept and pay for all of the lumber that is on grade.

If some of the lumber from a shipment has been used before the inspection is begun, what was used is considered to have been on grade. This "on grade" footage is taken into account when determining whether or not more than five percent of the total shipment is off grade.

A Moisture Claim

If a customer receives a shipment of lumber that appears to contain excessive moisture, a moisture claim must be filed within ten days of the lumber's arrival. Upon being notified of this potential problem, the trader must contact the mill immediately. The mill then has ten days in which to respond to the claim. If the customer notifies the trader of a moisture claim within three days of the arrival of the lumber, then the sawmill is obliged to also respond within three days.

Lumber with excessive moisture feels clammy, may have stain or fungus growing on it, and may feel particularly heavy. To check for moisture content, many customers will own, or have access to, a moisture meter.

A moisture meter is a hand-held device most commonly with two prongs on the front, an inch or so apart.

When the prongs are stuck into a piece of wood, an electrical current passes between them. The degree to which current flows between the prongs indicates the amount of moisture in the sample. A gauge on the face of the moisture meter shows the moisture reading. By sampling a number of pieces, a customer can quickly determine whether or not the lumber contains excessive moisture.

Upon notification of a possible moisture problem, the mill will check its records for recent dry kiln problems or other situations which could have resulted in inadequate drying of the lumber.

Because moisture content affects the weight of the wood, another means of determining excessive moisture content is by checking the weight per MBF of the entire shipment in question. The net weight of the lumber on a railcar or truck appears on the bill of lading. By dividing the net weight of the entire shipment by the total footage (the number of MBF) shipped, the actual weight per MBF is determined. By knowing the actual weights/MBF on a shipment, a comparison can be made with other recent shipments of the same item from that mill. If this load is found to be appreciably heavier than the average, it is a good indication that the lumber was not dry when it left the mill. If, on the other hand, the lumber is found to weigh about the same as the average weights for that mill and that item, the lumber most likely was dry when it left the mill.

This checking of weights only works if the car, truck, or pig-van carrying the lumber has been physically weighed or scaled. It will not work with railcars that ship on agreed weights.

It should also be noted that moisture content, and, indeed all grade characteristic, are based upon readings and grading at the mill, not upon arrival at the customer's yard. If lumber has traveled on an open flatcar or non-tarped truck and has been rained on, the surface moisture picked up is not the mill's responsibility.

How to handle a moisture claim. Moisture in lumber will evaporate if the lumber is simply left open and allowed to air dry. Stickering the pieces will speed up the process. If the lumber is needed immediately, it can be dried in a kiln. Most major cities around the country have custom dry kilns that will dry lumber for a charge. It must be remembered, though, that green lumber will shrink as it dries. If supposedly dry lumber contains a great deal of excess moisture when it is surfaced, once it becomes truly dry, it will be marginally smaller than regular surfaced dry lumber.

A Tally or Shortage Claim

A tally claim arises when the items shipped are different than what was ordered. A shortage claim arises when less lumber was shipped than was invoiced.

A tally claim usually happens when a mill's loading crew mis-loads some units of the wrong size or grade of lumber on an order. A tally claim can also happen if the mill runs short of a particular size and substitutes something else. Sometimes customers are tolerant and accept these minor changes in tally, other times the misshipment causes serious problems that result in a request for monetary compensation.

Shortage claims can be some of the most difficult to settle. On rail shipments, because the railroads require that railcars of lumber be fully packed and without gaps between units, mills can usually substantiate what was loaded on a particular railcar by referring to its *loading diagram*.

Shortage claims are frequently resolved when the customer realizes that a unit was miscounted or put in the wrong place during the unloading process.

The tricky part of a shortage claim is that until the missing lumber can be accounted for, the customer feels cheated by having to pay for something that was not received, and the mill feels justified in asking for payment on the total amount.

Shortage claims with trucks are a less frequent problem because the customer signs for what was received in the presence of the truck driver, who, in turn, had signed for the lumber when it was picked up at the mill. Any discrepancies can be easily traced.

How to handle a tally or shortage claim. When the tally received is not the tally that was ordered, and when this has been substantiated with photographs, a price adjustment is usually negotiated between the customer and the mill, and a price correction is made both on the mills and the customers invoice. If the mis-shipped item is something for which the customer has no use at any price, the trader may be obliged to help the mill sell that lumber somewhere else, at some renegotiated price.

In a shortage dispute, the customer needs to send a notarized letter stating the tally received. By comparing the mill's loading diagram with the customer's notarized tally, errors can frequently be picked up.

On a shortage claim, like with a moisture claim, net weights (from the freight bill) can also be used to help determine the amount of lumber on a shipment. To do this, divide the total net weight of the shipment by the footage the mill claims to have loaded. Then compare that weight with the mill's average weights for that item. If, by dividing the weight of the shipment by the mill's invoiced footage, the result is lower than the mills average per MBF weights, then there may well have been less lumber on the shipment than the mills invoice showed. Result showing higher than the mills average per MBF weights indicate that there was more lumber on the shipment than the mills invoice showed. If the result is approximately the same as the mills average, then the mill probably shipped the invoiced footage. As with moisture claims, this procedure does not work for a mill that ships on agreed weights.

A Shipment Time Claim

As mentioned in Chapter 4, shipment time refers to the date that lumber leaves the mill, unless it is otherwise specified on the order confirmation. Since the actual transit time is controlled completely by the carrier, a trader should never promise any particular arrival at destination time. Claims for too fast or too slow transit time seldom arise, and only in clearly negligent cases will carriers take responsibility for this.

A shipment time claim arises when a shipment leaves the mill either earlier or later than specified on the purchase order. An early shipment may create a problem if the customer has no place to store the material or is, say, taking a physical inventory and doesn't want the additional material on hand. In this type of situation, the orders to the customer and the mill should include the words "no sooner" next to the shipment time on the confirmation.

A late shipment (a much more common problem) may result in the customer's running out of lumber and being forced to purchase small quantities of lumber locally, invariably at a higher price. Industry practice holds that lumber is not seriously late as long as it leaves the mill within a week or so past expected shipment time, unless exact shipment time is discussed and stressed at the time of the order.

The actual date that an order leaves the mill can vary for any of several reasons: production behind in schedule, railcars or trucks that are ordered but don't arrive, foul weather, labor problems, etc. For these and other reasons, orders are usually written to give the mill some flexibility in shipment time. For example, shipping instructions such as, "two weeks or sooner,"or "thirty days or sooner," are quite normal.

How to handle a shipment time claim. If lumber ships early and arrives before it can be unloaded, it may have to either sit on the car or truck an extra day or two; or it may need to be unloaded at another site. Either of these alternatives will cost money, and claims might be filed for compensation. Another option when the shipment is early is to offer the customer extended terms of payment on the order. In this way, the customer has the material early, but does not need to pay for it until the expected arrival time.

If lumber happens to be late shipping from the mill to a customer in desperate need, compensation may have to be made to the customer for lumber purchased locally at higher cost. The customer may choose, and does have the right, to cancel a late shipping order entirely. This option is not frequently used, and then usually in a market where prices are falling and the customer can rebuy similar material at a lower price. In a falling market, a mill that is late on an order will occasionally renegotiate the sales price rather than have an order cancelled.

As a practical matter, customers are far more concerned with timely shipments in a falling market than in a rising market. (The buy price is fixed in either case, but in a falling market, unshipped lumber becomes increasingly less valuable--relative to replacement cost--as time goes on. In a falling market, customers want to get the lumber into their yards and get it sold before prices drop still farther.)

A Market Claim

A market claim is one in which the customer requests a downward adjustment in price because the market price has fallen since the time that the order was placed, even though the mill is not late in shipping. (A market claim is never received in a rising market!)

Although market claims occasionally come up, they are not considered legitimate or valid unless specific provisions are written into the order confirmation providing for this. Just as a mill will not seek an increased price if the market goes up before an order has shipped, a customer (or trader) cannot get a decreased price if the market goes down.

A market that has moved dramatically downward after an order is placed will tend to make customers more watchful for timely shipment and quality of stock than otherwise, but, generally, price adjustments are made only if the mill is out of contract.

GRIPES AND COMPLIMENTS

Gripes are not claims, but they deserve some attention here. Occasionally a customer will call to register a complaint about some feature of an order that is not serious enough to actually file a claim over. The sawmill should be told. This feedback is frequently the only way that a mill learns when its products are being badly received.

Likewise, when a customer calls in with a compliment, this, too, should be passed on to the mill. Mills, and traders, always hear about the problem orders; they seldom hear about the very good ones. Compliments and constructive comments are positive reinforcement which should be passed along.

An astute trader can frequently learn about a customer's likes and dislikes while handling a claim or a gripe. This information should be recorded to help put successful orders together in the future.

SUMMARY

Let's summarize the procedure one should follow when presented with a customer claim.

1. Remember Rule 1: Settle as expeditiously as possible.
2. Remember Rule 2: Document.
3. Get the material unloaded from the truck or car (unless the claim is a transportation claim, in which case the lumber must be inspected **before** unloading).

4. Have the material held intact, preferably under covered storage.
5. Verify by gradestamp, car or van number, and invoice, that the problem material is the right order.
6. Have the buyer personally inspect the lumber and give you an assessment of the type of problem, the severity of the problem, and an idea of, roughly, what an equitable settlement might be. Request moisture readings, pictures, notarized tallies, railroad inspector's report, etc., depending on the type of problem.
7. Make sure that payment is being made, at least on the acceptable portion of the material.
8. Contact the mill immediately. Pass along all pertinent information. Even if you don't know the actual severity of the problem, alert the mill so that it can be ready as the situation develops.
9. Several calls back and forth, to the customer and to the mill, may be required before everyone has a clear understanding of the problem. This must be done with as much speed as possible to keep within the time limits specified for that type of claim.
10. If, at the time all parties understand the problem, agreement still cannot be reached, a decision will have to be made to either: call for a reinspection, negotiated the best settlement possible (usually price adjustment), or have the lumber removed from the customer's yard (usually at the mill's or trader's expense) for resale elsewhere.

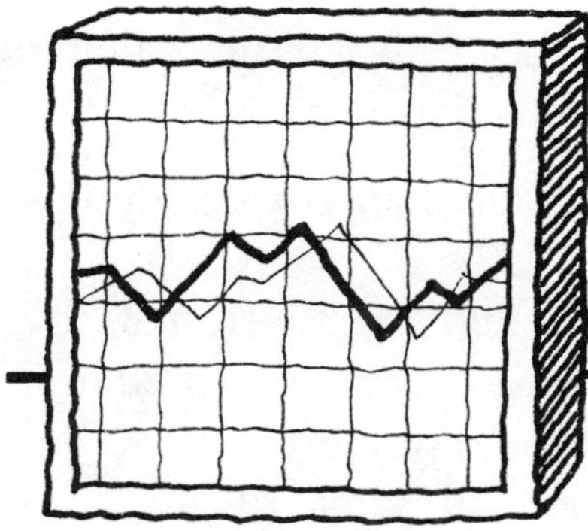

CHAPTER
SEVEN

The Futures Market

Occasionally buyers or sellers in the lumber industry would like to be able to firm price lumber with a shipment date farther into the future than the usual three to four weeks that most lumber is traded in. Some examples of this are building contractors who must bid the lumber for jobs that won't start for months, sawmills that periodically wish to sell lumber at current prices even though the lumber will not be produced until later, and lumber stocking yards that know they will need lumber in the future and would like to price commit (buy) at current market levels for part of those needs. To satisfy these and other demands that we shall explore, the lumber futures market was established. This chapter will look, in basic form, at what this market is and how it functions.

The lumber futures market was created, and is operated, by the Chicago Mercantile Exchange in Chicago,

Illinois. A lumber futures contract calls for the seller to provide, and the buyer to accept and pay for:

1. two carloads of 2x4 (a total of approximately 150 MBF);
2. a good random assortment of lengths from 8' to 16' or 20' (options for the tally are specified in the Mercantile Exchanges' literature);
3. lumber that is gradestamped "standard and better" or "2/Btr";
4. the species of SPF, Hem/Fir, or ES/LP;
5. lumber that is dry, S4S, and has eased edges; and
6. lumber that is loaded, paper-wrapped, onto a flat-car.

The one lumber item traded on the lumber futures market, dry R/L 2x4, is the single largest volume lumber item produced and traded in North America. As shown, the contract species are SPF, Hem/Fir, or ES/LP. These species are called *inland species* because, being spruce, pine, and White Fir, they grow east of the Cascade mountain range. (The primary *coastal species* are Douglas Fir and hemlock. These grow predominantly west of the Cascades.) For the last several years, SPF, which is a Canadian gradestamp only, has been the vastly predominant species shipped on lumber futures contracts. There are two primary reasons for this: freight considerations favor Canadian over Pacific Northwest production when shipping into the main consuming region (the southeastern quarter of the U.S.); and far greater quantities of SPF are produced than are either Hem-Fir or ES/LP.

A lumber futures contract is bought and sold like a stock, through a stock or commodity futures broker. Anyone having an account with one of these brokers may buy or sell a lumber futures contract. The buying or selling price is the trading price on the floor of the Exchange at the moment a contract is traded. Once an order (a contract) is placed, the price for that particular trade is fixed. The trading price on the floor of the exchange, however, continues to constantly change

When buying or selling a lumber futures contract, one is buying or selling a 2 C/L commitment in a particular predetermined contract shipment month. Contract shipment months are spaced 2 months apart, starting with January, subject to change by the Exchange. Trading in each contract shipment month goes on for 14 months prior to its expiration. In this way, there have been, and are currently, seven different contract months, or time frames, in which to buy or sell lumber futures contracts. Actual shipment of lumber occurs only after a contract has expired. (A contract expires on the 15th day of the contract expiration month.)

Once an order in a particular shipment month is either bought or sold, it need not be kept through to the contract expiration. To get out of an order commitment any time before contract expiration, one must place another order in the same contract month, but opposite to the earlier position. That is, the second order must be to sell a contract if the first order had been to buy a contract, or vice versa. In this way, the two trades offset one another, and the Exchange clears the buyer/seller from its records. The difference between the buying price and the selling price (plus the brokerage house fees) determines whether money is made or lost on the transaction.

In addition to this purpose of enabling people to buy and sell 2x4 for future shipments, the lumber futures market serves three other purposes. These are: 1) to enable the hedging of inventories or shorter-term *cash* 2x4 positions, 2) to provide opportunity for those wanting to trade lumber items other than 2x4s, and, 3) to enable raw market speculation. Here is a brief explanation of each of these other purposes.

Hedging

A hedge is like an insurance policy. Its purpose is protection from a major loss. A hedge is not placed to make additional profit; a hedge is used to protect a profit (or to minimize a loss) that already exists.

Cash, or the *cash market*, in futures jargon, is the real-world market that most lumber is traded in every day. The buying from sawmills, the selling to customers, and the negotiating over the phone on price and tally are all elements of the cash market. Everything discussed in Chapters 1-6 has dealt with the cash market.

Cash market lumber trading and the holding of cash market lumber inventories involves forecasting what prices are going to be in the future. Inevitably, some of these forecasts are going to be wrong. It is sometimes prudent to hedge a cash market position when lumber prices do not appear to be moving in the direction hoped for.

One can use the lumber futures market to hedge a cash market lumber position because the cash market and the futures market are related. If the price of 2x4 on the cash market goes up, the odds are that the price will also go up on the futures market. Similarly, if the price moves down on the cash market, it generally will move down on the futures market. Realizing this link, if one has a cash market 2x4 position, long or short, and the cash market price does not appear to be moving in the desired direction, one can hedge that position by creating an **opposite** position in the futures market.

For example, let's suppose that a trader has a long position in the cash market. That is, the trader owns 2x4 and thinks that prices will go higher. After being in the position for a short time, it becomes clear that instead of going up, prices may actually begin to drop. The trader is faced with two choices: either sell the cash 2x4 as quickly as possible at the best price possible (and that price may not be very good if panic selling sets in), or hedge the cash position on the futures market.

To hedge a long cash 2x4 position, the trader sells (goes short) a number of futures contracts to equal the footage of the long cash position. (Remember, each futures contract is for 150 MBF). Having done this, the trader is now both long (cash) and short (futures) approximately the same amount of board footage. If the cash market does fall, the futures market will likely fall also. The loss incurred when finally selling the cash market 2x4 will be offset by a profit realized when simultaneously buying back the futures market contracts. Thus, the hedge provides protection from serious loss. This scenario works the same way in reverse. If one is short on the cash market and prices start to climb, one buys futures contracts.

The real advantage of the futures market in these instances is speed. Tremendous quantities can be traded quickly on the futures market. Cash market trading is much slower, particularly in a falling market.

The principal force underlying the futures to cash relationship is something called *convergence*. Convergence is what happens at the expiration of a contract, when the theoretical lumber that has been traded for the previous 14 months becomes actual lumber that is going to ship from a sawmill. In the last several weeks before expiration, the trading price on the expiring month's futures contract and the trading price of R/L 2x4 on the cash market directly compete with each other, and so they tend to come together, or converge. This convergence, at the close of each contract, is what ties the cash market and the futures market together.

Basis Trading

Basis trading means taking advantage of the difference in price between the futures market and the cash market. To illustrate, let's say that in the middle of April, the cash R/L 2x4 market is trading at $160/MBF and the May expiration futures contract is trading at $175/MBF. The $15/MBF difference in price between the markets is called the basis. To take advantage of this basis, one should buy cash 2x4 for shipment "four weeks, no sooner" (meaning that it will ship in the middle of May) at $160/MBF and sell May futures lumber (which will expire and ship in the last half of May) at $175/MBF. By doing this, come the middle of May and the two markets converge, as one sells the cash market lumber and buys back the futures (before the contract closes), one should theoretically be able to make $15/MBF on the trade. And it doesn't matter whether the cash market rises to meet the futures market price (in which case, no profit is made on the futures market transaction, but

$15/MBF appreciation is realized on the cash trade), the futures market price drops to meet the cash price (so that no profit is realized on the cash trade, but $15/MBF is made when buying back the futures contracts), or if they meet in the middle (with some profit being made in each transaction).

Any item, including but not limited to R/L 2x4, can be basis traded. However, for a non-R/L 2x4 basis trade to be successful, there must be a high correlation in price movement between standard and better, R/L, 2x4 (the futures contract item), and the non-R/L 2x4 item being basis traded.

Basis trading non-R/L 2x4 assumes that since R/L 2x4 is the single largest lumber item produced and traded, as its' market price goes up or down, the prices on other lumber items will, to a greater or lesser extent, follow. To the degree that another lumber item does follow the price fluctuations of R/L 2x4, this other item may be successfully basis traded using the futures market.

For example, historically the price of dry, inland, R/L, 2x6, behaves very similarly to the price of dry, inland, R/L, 2x4. That is, there is a high correlation between movements in the price of R/L 2x4 and simultaneous movements in the price of R/L 2x6; the prices for these items tend to go up and down together. The opportunity, therefore, to successfully basis trade R/L 2x6 is good.

To illustrate, let us suppose that someone has 300 MBF of dry, inland, R/L, 2x6 in a warehouse, and the market price on that item starts to drop rapidly. Knowing the 2x4-to-2x6 price movement correlation, the owner of the R/L 2x6 could sell two futures market contracts (go short two contracts) and reasonably expect that as the cash market price on 2x6 falls (and the inventory loses value), so too will the futures 2x4 price fall. When the time comes to buy the two futures contracts back, the buying price will be lower than the earlier selling price. The profit made on the two futures contracts should approximately offset the loss incurred when the cash 2x6 is sold.

There is no guarantee, of course, that futures market 2x4 and cash market 2x6 will change price together in this instance. However, when faced with a guaranteed falling cash market, the odds are good that a 2x6 basis trade will be more profitable than doing nothing at all.

But not all lumber items behave pricewise the same way. That is, the prices of many lumber items, like (just as examples) 1x6 #3 Ponderosa Pine or 2x12 economy Douglas Fir do not move simultaneously with R/L 2x4. Therefore, the correlation between price movements of 1x6 #3 Ponderosa, or 2x12 economy Douglas Fir, and R/L 2x4 is not very high. The 1x6 and the 2x12 are used in applications that are entirely different from R/L 2x4. The supply and demand cycle for each of these items is driven by different factors, and it is only in the broadest view that these markets influence each other. For these reasons, the chances of profitably basis trading either of these items is not good

One can assess the likelihood of successfully using the futures market as a basis trading tool for a non-R/L 2x4 lumber item by historically charting price movements of that item and then comparing those movements to the historical price movement of dry, inland, R/L, 2x4. If the correlation is high, the odds of profitably basis trading the item is good.

Speculating

Some individuals are interested only in speculating on lumber futures prices. These people buy lumber futures contracts when they think that futures prices are going up, and they sell contracts when they think that prices are going to go down. Speculators rarely keep contracts through to expiration because they do not operate in the cash market. They have no cash market position to hedge, and they have no cash lumber commitments to sell or buy.

Speculators, nonetheless, have a legitimate role to play in the futures market. It is the speculators that supply additional liquidity to the market. They keep the market trading when other types of traders might only be willing to sell or only willing to buy. If everyone wants to buy, and no one wants to sell, literally nothing in the futures market can happen. Because speculators frequently have different criteria for buying and selling than do cash market traders, they enable the market to continue functioning.

APPENDIX A

Calculating Tally Value

When negotiating lumber tallies, particularly dimension lumber tallies, it is occasionally necessary to calculate the exact average price for a particular loading. To show how this is done, we will go back to the example introduced in Chapter 5. You, as the trader, are offering one mill's 2x4 tally, and the customer is requesting a different 2x4 tally. The tally that the mill is offering is: 2x4--2/8', 3/10', 6/12', 7/14', 8/16', 2/18', 1/20'. The customer has decided on a tally of: 2x4-- 2/8', 9/10', 1/12', 7/14', 8/16', 2/18', 1/20'.

The difference in value between the tallies can be calculated. To do this, first establish a price for each length. This is done by using the most current published prices, or using the mill's straight length prices. Let's suppose that you have checked both sources and have determined the prices are: 8': \$143/MBF, 10': \$161/MBF, 12': \$137/MBF, 14': \$140/MBF, 16': \$172/MBF, 18': \$154/MBF, and 20': \$157/MBF.

Next calculate the board footage, by length, on the two tallies (you know that the mill's unit size is 260 pieces per unit). To do this, you use the formula for calculating board footage, which is:

$$\frac{\text{Thickness (in.) x Width (in.) x Length (ft.)}}{12}$$

Using this formula, you see that an 8' long 2x4 is (2 x 4 x 8)/12, or 5.333 board feet. If you know that one piece of 8' long 2x4 contains 5.333 board feet, you can calculate the footage in an entire unit of 8' long pieces by multiplying 5.333 times the number of pieces in the unit (260 in our example.)

If you want to know the board footage in more than one unit, you simply multiply the board footage in one unit by the number of units in question. The following expanded formula makes tally computation for board footage per length easy.

$$\frac{T \times W \times L \times P \times U}{12}$$

T = Thickness (in inches)
W = Width (in inches)
L = Length (in feet)
P = Number of pieces per unit
U = Number of units

Remember this one formula. Board footage on any lumber item or items can be quickly calculated using it. Using the formula, for example, you find that the board footage in 8 units of 16' long 2x4 (260 pieces/unit) is (2 x 4 x 16 x 260 x 8)/12, or 22,187 board feet (22.187 MBF).

Table A-1 shows the calculated board footages, by length, for both the mills tally and your customers tally.

TABLE A-I

MILL'S TALLY		CUSTOMER'S TALLY	
2/8' =	2.773 MBF	2/8' =	2.773 MBF
3/10' =	5.200 MBF	9/10' =	15.600 MBF
6/12' =	12.480 MBF	1/12' =	2.080 MBF
7/14' =	16.987 MBF	7/14' =	16.987 MBF
8/16' =	22.187 MBF	8/16' =	22.187 MBF
2/18' =	6.240 MBF	2/18' =	6.240 MBF
1/20' =	3.467 MBF	1/20' =	3.467 MBF
Total =	69.334 MBF	Total =	69.334 MBF

Once the footage by length has been determined, multiply the value of each length by the footage of that length. This gives you a dollar value for that amount of footage on each car (See Table A-II).

TABLE A-II

MILL'S TALLY

Tally	Total Footage		Value of Length	Total Value
2/8' =	2.773 MBF	x	$143/MBF =	$396.54
3/10' =	5.200 MBF	x	$161/MBF =	$837.20
6/12' =	12.480 MBF	x	$137/MBF =	$1,709.76
7/14' =	16.987 MBF	x	$140/MBF =	$2,378.18
8/16' =	22.187 MBF	x	$172/MBF =	$3,816.16
2/18' =	6.240 MBF	x	$154/MBF =	$960.96
1/20' =	3.467 MBF	x	$157/MBF =	$544.32
Total	69.334 MBF			$10,643.12

(continued next page)

(Table A-II, continued)

CUSTOMER'S TALLY

Tally		Total Footage		Value of Length	Total Value
2/8'	=	2.773 MBF	x	$143/MBF =	$396.54
9/10'	=	15.600 MBF	x	$161/MBF =	$2,511.60
1/12'	=	2.080 MBF	x	$137/MBF =	$284.96
7/14'	=	16.987 MBF	x	$140/MBF =	$2,378.18
8/16'	=	22.187 MBF	x	$172/MBF =	$3,816.16
2/18'	=	6.240 MBF	x	$154/MBF =	$960.96
1/20'	=	3.467 MBF	x	$157/MBF =	$544.32
Total		69.334 MBF			$ 10,892.72

Finally, divide the total dollar value of each tally by the total number of board feet in each, and you come up with the average value per MBF of each tally:

You see that average value of the mills tally is:

$10,643.12/69.334 MBF = $153.50/MBF*.

The average value of the tally requested by your customer is: $10,892.72/69.334 MBF = $157.00/MBF*.

The difference in price between the tallies is $3.50/MBF.

* rounded to the nearest $.25/MBF

Glossary

(with page references)

Most of the words listed in the following glossary refer to terms from the text and are followed by a page reference. Some of the terms listed below do not define words from the text, but are commonly used in the industry and have been, therefore, included for their reference value.

AAC. After arrival of car. See page 87.

AAT. After arrival of truck. See page 87.

ADF. After deducting freight. See page 86.

ADI. After date of invoice. See page 87.

A-Frame flatcar. A railroad flatcar with a tall center divider down the length of the car. Sometimes called a centerbeam flatcar. See page 80.

Agreed weights. A freight term for the predetermined weights per MBF used for calculating total shipping weight on railroad invoices (rather than weighing each car). See page 33.

Air dried. Describes lumber that has been dried naturally, by the elements, as opposed to being dried in a kiln. See page 13.

All-door car. A railroad boxcar with fully-sliding side panels. See page 81.

Anti-stain treatment. A dipped or sprayed-on chemical treatment applied to lumber to prevent discoloration and the growth of fungus while it is in transit and storage. See page 74.

Anti-travel treatment. See anti-stain treatment (listed above), and page 74.

Association weights. Averaged lumber weights used to calculate freight by mills selling on a delivered price basis. See page 31.

Average weights. Sometimes called association weights. Used to calculate freight by mills selling lumber on a delivered price basis.

Back-to-back. A trading term referring to when a buy order and a sell order occur together. See page 17.

Bad ordered. A railcar that is removed from a train to fix mechanical or load problems. See page 83.

Barge. A transportation method used for hauling extremely large quantities of primarily commodity dimension lumber in ocean going vessels from west coast ports to east coast ports. See page 40.

Basis. A futures market term for the difference between the current futures market trading price (for a particular contract month) and the current cash market trading price. See page 128.

Basis trade. A cash market and futures market joint trade placed to take advantage of an existing futures market basis. See page 128.

BHFC. Bulkhead flatcar. See page 81.

Bill of lading. A written acknowledgment by a cargo carrier describing both the goods received for shipment and the instructions for shipment. See page 25.

Blocking & bracing. Low-grade lumber used in a railcar to keep the contents from shifting while in transit.

Boards. Generally refers to lumber 1" in thickness. See page 49, 53.

Board foot. A unit of measure for lumber, equivalent to a piece of wood 1" thick, 12" wide, and 1' long. Calculated using the following formula. See page 5.

$$\frac{\text{Thickness (in.)} \times \text{Width (in.)} \times \text{Length (ft.)}}{12}$$

Board measure. Means that the unit of measure being used is the board foot.

Bow (a lumber defect). A deviation from a straight line drawn end to end across the wide face of a piece of lumber.

Boxcar. A fully enclosed railcar with single or double sliding doors for access. See page 79.

Boxed heart. The pith, or center of the tree, enclosed within the four sides of a piece of lumber.

Broker. A term used to describe someone who brings a buyer and a seller together to consummate an order, then earns a commission based upon that order. See page 16.

Buck. To rough cut, for length, a log or long timber. See page 10.

Bulkhead flat. A flat deck railcar with a vertical wall built at each end. See page 80.

C/L. Carload.

Cant. A large flat-sided timber intended to be remanufactured to smaller sizes. See page 12.

Car number. The "license plate" of a railcar, painted on the sides and used for identification.

Carrier. A transportation term for the company contracted to haul lumber from one place to another. See page 24.

Cash discount. A percentage (usually) discount that may be deducted from an invoice if the invoice is paid promptly. See page 86.

Cash market or cash. A futures market term to describe the non-futures lumber market. See page 126.

Centerbeam flatcar. A railroad flatcar with a tall center divider built down the length of the car. Sometimes called an A-frame flatcar. See page 80.

Certificate of grade. A document issued by the head grader of a sawmill, certifying the grade and species classification of a particular load of lumber that has not been gradestamped. See page 68.

Checking (a lumber defect). A separation of wood fiber across the annual growth rings. See page 60.

Clean and bright. The freshly-manufactured appearance of lumber as it comes from the planer. See page 72.

Clear. Although a grade in itself, usually used as a generic term for select or finish grade lumber.

Coastal species. Species of trees growing west of the Cascade Mountains, the primary ones being Douglas Fir, Mountain Hemlock, Western Red Cedar, Redwood, and Sitka Spruce. See page 124.

Combination rate. A rail freight price quoted in two or more parts. See page 27.

Commodity. A lumber item of wide usage, traded in great volume.

Commons. The most generally used grade classification for boards. Sometimes used synonymously with boards. See page 53.

Concentration yard. A large lumber storage and reload facility.

Consignee. The one to whom goods are consigned or delivered; the owner of the goods being shipped.

Contract month. A futures-market term for the month in which a particular futures contract will expire. See page 125.

Contract rate. A preferential non-public freight rate negotiated between a shipper and a transportation company. See page 26.

Contract shipment month. A futures market term to describe the month in which a particular futures market contract will expire and ship. See page 125.

Convergence. A futures market term to describe the coming together of the futures-market price and the cash-market price at the expiration of a contract month. See page 127.

Count. The nominal thickness, width, and length used in determining board footage for invoicing purposes. See page 99.

Counter. A comeback, or counter offer to buy (sell) at a better price, tally, shipment time, etc., than was originally offered. See page 101.

Coverage. Both the ability and the willingness to quote a specific lumber inquiry. See page 103.

Crook (a lumber defect). A deviation from a straight line drawn end to end across the narrow face of a piece of lumber.

Cross-cut. To cut a piece of lumber perpendicular to its length.

Cup (a lumber defect). A deviation from a straight line drawn edge to edge across the face of a piece of lumber.

Cut. Sawmill terminology for the making of rough lumber. See page 10.

cwt. Hundredweight.

DET. Double end trimmed. See page 66.

Debarker. A sawmill machine that removes the bark from a log. See page 11.

Delivered price. A price delivered to a particular destination that includes the cost of the lumber as well as the freight. See page 29.

Delivering carrier. The railroad serving the destination customer. See page 24.

Demurrage. A fee assessed by railroads to shippers who take extra time with cars beyond the normally allotted loading, unloading, and transit time. See page 20.

Dense. A term used to describe the specific gravity (i.e., strength) of the wood fibers in lumber as measured by the number of rings per inch and the percentage of summerwood to springwood. Dense lumber (usually measured only in Douglas Fir and Larch) must contain 6 or more annual rings per inch plus 1/3 or more summerwood. There are exceptions; see the gradebook.

Design values. Strength factors of lumber based upon six different properties or methods of assessment. These six methods of assessment measure lumber for its fiber stress in bending (Fb), tension parallel to grain (Ft), horizontal shear (Fv), compression parallel to grain (Fc), compression perpendicular to grain (Fc), and modulus of elasticity (E).

DF-L. Douglas Fir and Larch.

Dimension lumber. Lumber usually two inches in thickness. See page 48.

Distribution yard. A large metropolitan storage and shipment yard stocking lumber and other building materials. See page 9.

Diversion. The changing of the destination of a railroad car while it is in transit. See page 20.

Diversion point. A centrally-located geographic point to which unsold railcars of lumber are shipped until the lumber is sold and the cars can be diverted to the final destination. See page 20.

Door-to-door. A piggyback flat or van rate that includes service from the point where the lumber has been loaded in a piggyback flat or van at the mill, and goes through to delivery in the same flat or van to the customers yard. See page 38.

Double end trimmed (DET). Lumber that is trimmed reasonably square by a saw on both ends. See page 66.

Dressed. Clean surfaced. See page 63.

Dry. Lumber with a moisture content of less than 19% (less than 15% for selects and factory lumber). See page 61.

Dry kiln. A special room or building in which lumber is dried by controlling the moisture and temperature. See page 61.

Dunnage. Low-grade lumber that is used as blocking and bracing when loading lumber onto railroad, truck, or van equipment.

Eased edge. A slight rounding of the edges of lumber, done as the lumber goes through the planer. See page 66.

Edgebranding. An inked marking, usually indicating the manufacturer, printed continuously down the narrow face of a piece of lumber.

Edger. A piece of sawmill machinery that cuts wide, irregularly-edged lumber, into narrower uniformly-edged lumber. See page 12.

End matched. A pattern run on both ends of a piece of lumber; one end is given a concave shape, one is given a convex shape. When two end matched pieces of lumber are butted end-to-end, the convex pattern of one piece fits into the concave pattern of the next, forming a joint.

End paint. A colored paint sprayed onto the ends of lumber both to help prevent the ends from splitting and as a means of manufacturer identification. See page 14.

End stamp. A gradestamp placed on the end of finished lumber, instead of the more normal placement on the face. See page 68.

End trim. The description of how the ends of finished lumber are cut. See page 66.

End wax. Wax sprayed onto the ends of lumber to help prevent the ends from splitting. See page 14.

Engineering design values. Strength ratings for different characteristics of lumber. Used by builders, architects, and engineers to determine proper sizing of lumber when designing structures.

ES/LP. Engelmann Spruce and Lodgepole Pine.

Excess rate. A reduced rate charged on railcar shipments for all weight in excess of the minimum weight. See page 27.

FC/PUTT. Flatcar, poly under top tier.

FC/PW. Flatcar, paper wrapped.

Face. Usually refers to the best of the two wide sides of a piece of lumber. See page 52.

Factory lumber. Lumber intended to be remanufactured after it leaves the sawmill. See page 58.

FG. See Flat grain.

Finger jointed. A method of end joining lumber whereby the ends are cut in a deep zigzag pattern forming "fingers." Using glue, heat, and mechanical pressure, the pieces are forced together to form a strong end-to-end joint.

Finish (selects). The highest grades of lumber graded under Alternate Board Rules. See page 53.

Firm offer. A promise to buy or sell a product if certain specific conditions are met. See page 101.

Firm price bid. A fixed-price quotation on lumber that is not to be shipped for some extended period of time.

First factor. A freight term referring to the first part of a split rate. The first part, or first factor, is the rate for the minimum weight requirement, the second part covers the excess. See page 34.

Flatcar. A railcar open on all sides. A platform on wheels. See page 80.

Flat deck truck. A truck tractor and trailer designed for over-the-road hauling. The trailer is a flat platform on wheels from which unitized lumber (or other things) can be loaded and unloaded by forklift. See page 78.

Flat grain. Lumber sawn so that the wide face of the finished piece is approximately parallel (less than 45°) to the annual growth rings of the log.

Flat-to-van. A piggyback van rate characterized by service from the point where the lumber is picked up on a flat-deck truck at the mill, to the point where it is delivered by van to the customers yard. See page 39.

FOB mill. See "Free on board" and page 29.

FOHC. Free of heart center (boxed heart).

Free on board. A freight term meaning that material being purchased is going to be loaded on the conveyance chosen to ship it, as part of the purchase price. See page 29.

Freight collect. A freight bill intended to be paid upon arrival at the destination. See page 25.

Freight prepaid. A freight bill that is paid at the time of shipment, rather than once the shipment has arrived at the destination. See page 25.

Framing lumber. Usually refers to 2" thick (dimension) lumber which is 4" to 12" wide and is used in building applications such as studs, joists, and rafters.

Fresh run. Lumber that has been recently surfaced. See page 72.

Full sawn. Lumber that is cut, in the rough, to its full nominal size. See page 65

Furring strips. 1x2, 1x3, and 1x4 of middle to low grade lumber that are used in general building applications.

Gang-rip. The cutting of one piece of lumber in numerous smaller pieces, in just one pass through the saws. See page 65.

Gang saw. A piece of sawmill equipment that cuts lumber, in one pass, into a number of smaller pieces. See page 12.

Grademark. The crayon mark made on a piece of lumber, by the grader at the sawmill, to designate grade.

Gradestamp. An inked marking put on lumber to show grade and other important characteristics of the piece. See page 68.

Green. Lumber with a moisture content in excess of 19%. See page 61.

Green chain. A moving conveyor chain bringing rough, green lumber out of the sawmill. From the green chain, lumber is sorted by thickness and width in preparation for drying or surfacing.

H & M. See "Hit and Miss."

H or M. See "Hit or Miss."

Headrig. The primary saw in a sawmill. See page 11.

Heartwood. The central portion of a log between the center (pith) and the sapwood.

Hedge. A transaction intended to protect the existing profit, or minimize the existing loss, of an earlier transaction. See page 126.

Hem-Fir. A species grouping including western hemlock and all true firs except alpine and balsam. See page 47.

Hit and miss. A term that describes the surfacing on lumber containing a series of skips. The skips, or skip dressing (meaning the unsurfaced portions) may not be over 1/16" deep and must have surfaced areas between them. See page 99.

Hit or miss. A term that describes the surfacing on lumber that is completely surfaced, partly surfaced, or entirely rough. Where rough, the scantness may be 1/16" deep.

Hundredweight. One hundred pounds. Abbreviated "cwt." See page 27.

Incentive rate. A class of low freight rates based upon a high minimum weight, designed to encourage the loading of more lumber onto transportation equipment. See page 27.

Inland species. Species of trees growing east of the cascade mountains, the primary ones being most spruce, most pine and most true firs. See page 124.

Interchange (R/R). A point at which different railroads intersect and railcars can be switched from one carrier to another. See page 25.

Intermediate carrier. A railroad in a routing that is neither the originating carrier nor the delivering carrier. See page 25.

Joist. A structural framing member used horizontally to support a ceiling or floor.

Kerf. The gap or space left by a saw as it cuts a piece of lumber in two. See page 64.

Kiln. See dry kiln and page 61.

Kiln dried. Describes lumber that has been dried in a kiln (as opposed to being air dried). See page 13, 61.

Kiln wet. Describes lumber that has been through the kiln drying process, but still contains more moisture than is allowable for dry lumber. See page 62.

Loaded miles. A truck transportation term used for calculating truck freight rates, and referring to the actual number of miles that lumber will be carried from a mill to a destination. Does not include *dead-head*, or non-loaded, miles a trucker must travel to get to a mill to pick up the lumber. See page 35.

Loading. The type of equipment used to transport a shipment of lumber. See page 78.

Loading diagram. A freight term describing a diagram showing the placement and configuration of units of lumber in a loading. See page 117.

Load shift. Lumber units or pieces that have changed position while in transit, possibly damaging the material and making it more difficult to unload. See page 110.

Log deck. A holding yard for logs before they go into the sawmill. See page 11.

Long (position). A speculative market position in which one buys and holds lumber for some time before selling it. See page 19.

Loose loaded. A boxcar loading in which the lumber is not unitized, but instead is loaded and unloaded piece by piece. See page 79.

Lot loaded. A way of loading lumber on a railcar or truck that is going to make two or more destination stops, with each stop getting a percentage of the various items on the load.

Lumber. Wood that has been manufactured from a log and has been sawn on all four sides. See page 4.

Machine Stress Rated. Lumber grading that involves nondestructive mechanical stress testing and marking of the modulus of elasticity.

MBF. Thousand board feet. See page 5.

MC 15. Moisture content 15% or less. See page 62.

MG. See Mixed grain.

Mill run. The general grade description for a mill's entire grade tally on a particular run, with nothing having been selected out.

Minimum weight. A freight term meaning the weight that will be charged for, regardless of whether that much weight actually ships. See page 26.

Minor bundled. Sometimes called *sub-bundled.* Describes special packaging for smaller pieces of lumber (all the same size) that are tied together with string or tape, twelve to twenty-five per bundle. These small bundles are then usually unitized into normal-sized (approx 2' x 4') units with steel banding.

Mismanufactured. Lumber that is defective due to errors in its milling or planing. See page 112.

Mixed grain. A load of lumber that will contain pieces of both flat grain and vertical grain.

Mixed species. A species grouping that may include any western softwood. See page 45.

Moisture content. The weight of the water in wood, expressed as a percentage of the weight of the oven dry (containing no moisture) wood. See page 59, 114.

Moisture meter. An instrument used to determine the moisture content of a piece of wood. See page 114.

Modulus of elasticity. A ratio describing the amount a material will deflect in proportion to an applied load.

MSR. Machine stress rated. See page 57.

Nested shorts. Short lengths of lumber that are placed end to end, inside units of longer length lumber. See page 72.

Net-net. An invoicing term meaning that no cash discount is given for prompt payment. These are the invoicing terms used in all futures-market transactions.

NLGA. National Lumber Grades Authority.

Nominal. The name for the size of a piece of lumber, not the actual measurement of that piece of lumber (i.e., what we call a 2x4 is actually a piece of lumber measuring 1-1/2" x 3-1/2"). See page 49.

O/H. See On hand, and page 20.

Off grade. Lumber that does not conform to the grade it was represented as being. See page 105.

Office wholesaler. Usually used as a synonym for a lumber trader. See page 7.

Off the market. A trading term meaning that a buyer or seller is not currently soliciting business.

On hand. A rolling railcar that has stopped at a diversion point pending further orders from the shipper. See page 20.

Open units. Lumber that is not protected by paper wrap or PUTT. See page 75.

Order file. Refers to the current unshipped orders of a sawmill.

Origination carrier. The railroad serving the sawmill from which a railcar is shipped. See page 24.

Oven dry. Lumber from which all the moisture has evaporated and so contains a 0% moisture content. See page 60.

PAD. See Partially air dried and page 62.

Paper wrapped. Heavy paper wrapped around the top and four sides of a unit of lumber, to protect it during transit and outside storage. See page 74.

Pattern. A series of grooves and channels incised down the length of a piece of lumber as it goes through a planer. See page 65.

Payment terms. See page 86.

Pencil trim. An invoicing procedure whereby less board footage is invoiced than is actually shipped. See page 66.

Per car charge. A freight term meaning a flat fee charged for hauling a railcar from one place to another, regardless of the footage or weight being shipped. See page 28.

PET. Precision end trimmed. See page 66.

Piggyback van. A van enclosed on the sides, the front, and the top, which is loaded and unloaded from the rear and is designed to be hauled either over the road or on a railroad flatcar. Also called a pig van. See page 38, 78.

Piggyback flat. A flat-deck truck trailer designed to be hauled either over the road or on a railroad flatcar. Also called a pig flat. See page 79.

Pith. The heart or bull's-eye of a tree. Seen in cross-section, at the center of the annual growth rings.

Poly under the top tier (PUTT). A way of giving some protection to unitized lumber by the insertion of a piece of plastic between the next-to-top and the top tiers. See page 74.

Planer. A piece of sawmill equipment that planes the sides of rough lumber leaving it smooth and uniform in size. See page 13.

Planer mill. Although sometimes a separate facility, the planer mill is usually that part of a sawmill where lumber is planed, graded, and sorted. See page 10.

PLIB. Pacific Lumber Inspection Bureau.

PP-LP. Ponderosa Pine and Lodgepole Pine.

PP-SP. Ponderosa Pine and Sugar Pine.

Precision end trimmed (PET). Lumber trimmed smooth on both ends and varying no more than 1/16" in no more than 20% of the pieces. See page 66.

Priced at time of shipment. An order that does not have a firm buy or sell price until the lumber is ready for shipment.

Prepay (the). A preferential US-Canadian dollar exchange rate allowed to shippers originating shipment in Canada with destinations in the United States, initiated by Canadian rail companies to encourage payment of freight bills at the time of shipment rather than arrival at destination. See page 32.

PTS. See Priced at time of shipment.

Published tariff. A freight term for the standard, printed freight rate from an origin to a destination. See page 26.

Pulled to length. Unitized lumber that contains only one length per package. See page 72.

PUTT. See Poly under the top tier and page 74.

Quote. The price, tally, shipment time, etc., coverage for an inquiry on a particular load of lumber. See page 103.

Rafter. Dimension lumber that provides framework for the roofing material in building construction.

Ramp-to-door. A piggyback flat or van rate characterized by service from the point where the lumber has been loaded in the van or flat, and placed at the railroad flatcar loading dock, on through to delivery to the customers yard. See page 40.

Random. Containing pieces of different size.

Random length. A lumber loading containing most or all of the different normal lengths for that item. See page 6, 72, 94.

Reciprocal switch. A railroad term referring to an agreement between railroad companies serving in the same town, to switch each others' cars for no additional switching fee. See page 84.

Reinspection. The process of manually regrading a shipment of lumber believed by a customer to be off grade. See page 107, 113.

Reload center. A large lumber storage and trans-shipment facility. See page 8.

Re-man. The process of remanufacturing lumber; or a remanufacturing facility.

Rerun (to pattern). Returning surfaced lumber to the planer to add a pattern. See page 65.

Resaw. The process of cutting lumber in two, parallel to its wide face. For example, making a 2x12 into 2 pieces of 1x12. See page 63.

Rip. The process of cutting lumber in two, perpendicular to its wide face. For example, making a 2x12 into 2 pieces of 2x6. See page 63.

Ripped and bundled. A remanufacturing process whereby lumber is ripped and then minor bundled.

RIS. Redwood Inspection Service.

R/L. See Random length.

Roller. An unsold railcar of lumber that is loaded and moving toward a diversion point. See page 20.

Rough. Lumber that has not been planed or surfaced. See page 65.

Routing. A freight term referring to a railcar's itinerary from origin to destination, including interchanges and intermediate carriers. See page 25.

Rule tally. A method of tallying, individually, the board footage of pieces of lumber that are not cut or surfaced to standard width sizes. Usually used on factory lumber that is surfaced S2S.

Run. Sawmill terminology for planing rough lumber. See page 10.

S1S2E. Surfaced one side, two edges. See page 64.

S2S. Surfaced two sides. See page 63.

S4S. Surfaced four sides. See page 63.

S2S1E. Surfaced two sides, one edge. See page 64.

Sapwood. The part of a log between the heartwood and the bark.

Saw sized. Lumber which is cut in the sawmill to finished surfaced sizes. In this way, it is interchangeable- size-wise- with surfaced lumber, without having to go through the planer. See page 65.

Scaled. A transportation term meaning weighed. See page 33.

Scant. Lumber (usually mismanufactured to) smaller than standard size.

Select. A high grade of lumber that allows few knots or defects. See page 53.

Shake (defect). A separation of wood fiber through the annual growth rings.

Shipment time. The time indicated for a shipment of lumber to leave from a sawmill. See page 81.

Shipper. A freight term for the party paying the freight bill. See page 26.

Short. A speculative position in which one sells lumber days or weeks before it is bought. See page 19.

Side cut. The outside cut of a log that is curved on one side and flat on the other. Usually remanufactured into 1x3, 1x4 or 1x6.

Siding (railroad). A short dead-end track coming off a main rail line, on which railcars are loaded and unloaded. See page 24.

Skip dressing. Areas on a piece of lumber that fail to surface smoothly, usually due to lumber which has been sawn scant in the sawmill.

Slope of grain. The angle formed by the intersection of the wood fibers with the side of a piece of wood.

Softwood. The wood from trees commonly referred to as conifers or evergreens. See page 3.

Species grouping. Different species of wood that are marketed together because they are harvested together and have similar characteristics that make them interchangeable in use. See page 45.

SPF. A Canadian species grouping that includes several varieties of spruce, pine, and fir. See page 47.

Splitterhead ripped. Lumber that is ripped not with a saw but instead on a planer or moulder. This produces a ripped face that is smoother and cleaner than that usually produced by a saw.

Springwood. Wood fiber that is produced during the spring of the year when moisture is plentiful and growth is rapid. When a log is viewed in cross section, the springwood is the light area between the dark rings.

Square edge. Lumber on which there is no wane and the edges have not been eased. See page 66.

Standard sawn. Lumber that is rough-cut to smaller than full-sawn size, yet large enough so that it still may be run through the planer and clean surfaced on all four sides. See page 65.

Stickers. Spacers placed between the tiers of rough green lumber to aid air circulation as the lumber dries. See page 12.

Stopover. A load of lumber that is picked up or dropped off at two or more points.

Stud. A piece of dimension lumber, usually PET, used as an upright in wall construction.

Stumpage. Timber not yet harvested.

Subject to prior sale. A trading phrase used to indicate that lumber is being offered for sale on the condition that no one else has already purchased it. See page 102.

Sub prior. An abbreviated term for "subject to prior sale." See page 102.

Summerwood. Wood fiber that is produced later in the year than the springwood. Because moisture is generally not as plentiful, growth is less rapid than during the spring. When a log is viewed in cross section, the summerwood is the dark rings.

Surfaced. Lumber that has gone through a planer so that its sides are smooth and uniform in size. See page 63.

Surface measure. A method used for measuring lumber that considers the face area only. For example, a 2x4 12' and a 1x4 12' both contain the same board footage (4 bd. ft.) when measured on a surface measure count.

Surface moisture. Rain or condensation moisture on the outside of otherwise dry lumber. Surface moisture will evaporate fairly quickly and not permanently alter the dry condition of the lumber.

Switch. A railroad term meaning to change a railcar from one carrier to another. See page 20, 84.

T/L. Truckload.

Tally. The thickness, width, and length piece count of lumber shipping on a particular order. See page 76, 93.

Team track. A public rail loading and unloading facility. See page 24.

Thousand board feet. The unit of measure most frequently used in the wholesale lumber industry, representing 1,000 board feet. See page 5.

Tier. A horizontal course in a unit of lumber. See page 74.

Timbers. Lumber 5" or more in thickness. See page 52.

Trace. To learn, from the carrier, the location of a load of lumber in transit from the mill to the destination. See page 83.

Trading (lumber). The buying, selling, and transporting of wholesale quantities of lumber. See page 16.

Transit. A term used in the wholesale lumber industry to describe unsold lumber that may or may not have actually shipped from a mill. See page 20.

Twist (defect). A deviation, flatwise, in a piece of lumber creating the form of a curl or a spiral.

UDDB. Unitized double door boxcar. See page 79.

Underweights. The difference between the actual weight of lumber and the association weight of lumber. Used in determining the freight to a destination, by mills that invoice on a delivered price basis. See page 31.

Unitized. Lumber that is steel banded together into units for forklift handling. See page 14.

Unit size. The number of pieces of lumber in a unit. See page 69.

Unseasoned. Lumber that contains more than 19% moisture content. See page 61.

Upcharge. An extra charge for some additional service.

Van-to-flat. A piggyback van rate characterized by service from the point where the lumber has been loaded in a van at the mill, on through to the point where it is delivered on flat-deck truck to the customers yard. See page 39.

VG. See Vertical grain.

Vertical grain. Lumber sawn so that the wide face of the finished piece is approximately perpendicular (greater than 45 degrees) to the annual growth rings of the log.

Wane (defect). The absence of wood along an edge or corner.

Warp (defect). Any combination of bow, crook, cup, or twist.

WCLIB. West Coast Lumber Inspection Bureau.

West coast wood. A species grouping including: Douglas fir, western larch, all true firs, all true spruce, and all true hemlock. See page 48.

Western. As used in this text, that portion of North America (the US and Canada) west of the Rocky Mountains. See page 4.

Western cedar. A species grouping including: Incense Cedar, Western Red Cedar, Port Orford Cedar, and Alaska Yellow Cedar. See page 48.

White woods. A species grouping including: Engelmann spruce, all true firs, all hemlocks, all pines. See page 48.

Wholesaling. A term frequently used synonymously with trading, refering to the buying, selling, and transporting of carload or truckload quantities of lumber. See page 16.

WWPA. Western Wood Products Association.

Study Questions and Answers

Chapter 1

1. Define the following terms:

    a. Softwood
    b. Lumber
    c. Board Foot
    d. MBF
    e. random length
    f. reload center
    g. distribution yard
    h. headrig
    i. planer

2. List 6 major Western softwood lumber species.

3. Calculate the number of board feet in each of the following:

    a. 1 piece 2x4--10 foot
    b. 1 piece 4x6--24 foot
    c. 1 piece 1x12--16 foot
    d. 6 pieces 2x8--12 foot
    e. 100 pieces 1x10--14 foot

f. 1 unit (294 pieces/unit) 2x4--8 foot
g. 3 units (128 pieces/unit) 1x12--20 foot
h. 8 units (105 pieces/unit) 2x10--18 foot

4. List the following procedures, first to last, as they occur in the chronological progression of dry surfaced lumber through a sawmill.

_____ a. drying
_____ b. at the headrig
_____ c. end trimming
_____ d. felling
_____ e. paper wrapping
_____ f. grading
_____ g. debarking
_____ h. through the edger
_____ i. unitizing
_____ j. stickering
_____ k. bucking
_____ l. planing

Chapter 2

1. Define the following terms:

a. lumber trading
b. a back-to back order
c. a short order
d. a long order

e. a roller
f. on hand
g. demurrage

Chapter 3

1. Define the following terms:

   a. railroad siding
   b. originating carrier
   c. delivering carrier
   d. bill of lading
   e. routing
   f. combination rate
   g. a "per car" charge
   h. FOB mill
   i. "the prepay" (Canadian)
   j. contract rates
   k. agreed weights
   l. association weights
   m. door-to-door van rate
   n. flat-to-van rate

2. Calculate the following freight rates in terms of a price/MBF.

   a. Spokane, WA to Memphis, TN
      equipment: 50' boxcar
      footage: 49.298 MBF
      lumber weight: 1750 lbs/ MBF
      rate: 3.28 on 85, 3.08 on excess

b. Columbia Falls, MT to Greenville, SC
equipment: 56' 8" flatcar
footage: 72 MBF
lumber weight: 1725 lbs/MBF
rate: 4.19 on 110, 3.97 on excess

c. Prince George, BC to Miami, FL
equipment: 50' flatcar
footage: 77.5 MBF
lumber weight: Mill ships on association weights of 1800 lbs/MBF
rate: 5.13 on 85

d. Eugene, OR to Savannah, GA
equipment: 60' BHFC
footage: 56 MBF
lumber weight: 3200 lbs/MBF
rate: 4.37 on 150, 3.67 on excess

e. Klamath Falls, OR to Norfolk, VA
equipment: 52' boxcar
footage: 54 MBF
lumber weight: 1750 lbs/MBF
rate: a) to Chicago-- 3.37 on 85, 3.17 on excess
b) beyond-- $1276/car

f. Post Falls, ID to Bowling Green, KY
equipment: flatdeck truck
footage: 28 MBF
lumber weight: 1700 lbs/MBF
rate: 4.82 on 48,000 lbs. minimum

g. Republic, WA to Cleveland, OH
   equipment: flat-to-van
   footage: 27 MBF
   lumber weight: 1766 lbs/MBF
   rate: $2435

h. Williams Lake, BC to Birmingham, AL
   equipment: door-to-door van
   footage: 29 MBF
   lumber weight: 1675/MBF
   rate: $2910 (US funds)

Chapter 4

1. Define the following terms:

   a. dry lumber
   b. PAD lumber
   c. kiln wet lumber
   d. S4S
   e. resawn lumber
   f. ripped lumber
   g. eased edges
   h. PET
   i. pencil trimmed
   j. DET
   k. pulled to length
   l. PUTT
   m. All-door car
   n. A-frame car

o. reciprocal switch
p. ADI (invoicing)
q. AAC (invoicing)

2. Why is there a difference between the sizes of surfaced green lumber, and that of surfaced dry lumber?

3. What is the difference between S1S2E and S2S1E, and describe how each is produced.

4. Why is shipment time quoted in terms of mill shipment rather than arrival at destination time?

5. List the allowable species in the following mixed species groupings:

   a. SPF
   b. White Woods
   c. Hem-Fir
   d. Western Cedar
   e. Fir/Larch

Chapter 5

1. Define the following terms:

   a. firm offer
   b. subject to prior sale

Chapter 6

1. Define the following terms:

    a. reinspection
    b. bad ordered

2. Describe briefly, likely cause for each of the following types of claims.

    a. transportation
    b. general grade or manufacturing
    c. moisture content
    d. tally or shortage
    e. shipment time

3. Discuss how you would resolve each of the five claims listed above.

Chapter 7

1. Define the following terms:

    a. hedge
    b. cash market
    c. convergence

2. Fill in the blanks

a. Each lumber futures contract represents the quantity of _________ MBF of lumber.

b. The lumber item and grade traded on the futures market is ___________.

c. The allowable mixed species groupings for shipment on lumber futures contracts are

_______________________________________.

d. Circle the allowable characteristics of lumber as defined by the Chicago Merchantile Exchange's 2x4 R/L lumber contract.

rough	R/L 8'-24'	paper wrapped
ES/LP	S-Green	loaded boxcar
PUTT	75 MBF	square edged
PET	Hem/Fir	Utility & Btr
S1S2E	150 MBF	Doug Fir/ Larch
SPF	2 & Btr	gradestamped
PAD	R/L 8'-20'	loaded flatcar
S4S	Doug Fir	eased edge
DET	R/L 6'-16'	Stand & Btr
S-Dry	S2S1E	White Woods

Appendix

1. Calculate the board footage for the following R/L 2x4 tally, assuming a unit size of 294/unit.

3/8', 5/10', 4/12', 9/14', 6/16', 2/18', 1/20'.

2. Using the same tally, calculate the average price/MBF, given the following prices per length.

 8' -$154/MBF
10' -$174/MBF
12' -$150/MBF
14' -$145/MBF
16' -$192/MBF
18' -$186/MBF
20' -$189/MBF

3. A customer calls and requests a price on a van or truckload of paper wrapped D & Btr Fir/Larch selects with a tally of 5-6 MBF 1x4, 7-8 MBF 1X6, 15-16 MBF 1x10. What will you quote?

Before you can quote the customer a price, you must get answers to each of the following questions from each different mill that has this particular specification of lumber available for sale.

1. What is the mill price for the lumber from mill A?
2. What are the van and truck rates to the destination from mill A?
3. Does mill A upcharge for loading vans? for loading trucks?
4. What does mill A charge for paper wrapping?
5. What does mill A think that their lumber will weigh (so that you can determine how much footage will fit on the truck/van)?

After compiling the information from mills A, B, C, and D, you find that mill C has the best overall delivered prices. This is the pricing information on mill C.

The FOB mill price on the 1x4 is $480/MBF, the 1x6 is $560/MBF, and the 1x10 is $540/MBF.

The van rate to your destination (door-to-door) is $2350, and that the mill will upcharge $5/MBF to load the van. The truck rate to the destination is $5.83/48, with no upcharge for truck loading.

The mill thinks that its' dry S4S Fir/Larch will weigh approximately 2050 lbs/MBF.

The mills upcharge for paper wrapping is $6/MBF.

If you want to make $20/MBF profit on this order, what are your delivered sell price quotations going to be?

ANSWERS TO STUDY QUESTIONS

Chapter 1

3. a. 6.66 bd ft
   b. 48.0 bd ft
   c. 16.0 bd ft
   d. 96.0 bd ft
   e. 1,166.6 bd ft (1.167 MBF)
   h. 1,568 bd ft (1.568 MBF)
   g. 7,680 bd ft (7.68 MBF)
   h. 25,200 bd ft (25.2 MBF)

4. d, k, g, b, h, j, a, l, f, c, i, e

Chapter 3

2.a. 49.298 MBF x 1750 lbs./MBF = 86,272 lbs., so,
85,000 lbs. @ $3.28/cwt = $2,788.00
1,272 lbs. @ $3.08/cwt = $ 39.18
Total Freight = $2,827.18

$2,827.18/49.298 MBF = $57.25/MBF*

(* rounded to the nearest $.25/MBF)

b. 72 MBF x 1725 lbs./MBF = 124,200 lbs., so,
110,000 lbs. @ \$4.19/cwt = \$4609.00
14,200 lbs. @ \$3.97/cwt = \$ 563.74
Total Freight = \$5172.74

\$5,172.74/72 MBF = \$71.75/MBF*

c. 1800 lbs./MBF x \$5.15/cwt = \$92.75/MBF

d. 56 MBF x 3200 lbs./MBF = 179,200 lbs., so,
150,000 lbs. @ \$4.37/cwt = \$6,555.00
29,200 lbs. @ \$3.67/cwt = \$1,071.64
Total Freight = \$7,626.64

\$7,626.64/56 MBF = \$136.25/MBF*

e. 54 MBF x 1750 lbs./MBF = 94,500 lbs., so,
85,000 lbs. @ \$3.37/cwt = \$2,864.50
9,500 lbs. @ \$3.17/cwt = \$ 301.15
Freight to Chicago = \$3,165.65

Freight to Chicago = \$3,165.65
plus beyond to Norfolk = \$1,276.00
Total Freight = \$4,441.65

\$4,441.65/54 MBF = \$82.25/MBF*

(* rounded to nearest \$.25/MBF)

f. 28 MBF x 1700 lbs./MBF = 47,600 lbs., so,
48,000 lbs. (min. rate) x $4.82/cwt = $2,313.60

$2,313.60/28 MBF = $82.75/MBF*

g. $2435/27 MBF = $90.25/MBF*

h. $2910/29 MBF = $100.25/MBF*

Chapter 4

2. Green lumber is surfaced to larger sizes than is surfaced dry lumber because green lumber will shrink to about the size of surfaced dry lumber as it air dries.

3. S1S2E is the product of an S4S piece of lumber that has been sawn in half parallel to the wide face. Two pieces of 1x12 from a piece of 2x12 is an example.

S2S1E is the product of as S4S piece of lumber that has bee sawn in half perpendicular to the wide face. Two pieces of 2x2 from a piece of 2x4 is an example.

4. Lumber shipment time is quoted in terms of mill shipment time rather than delivery to destination time, because neither the producing mill nor the selling traded can control the speed of a shipment once it has been loaded onto the transportation conveyance at the mill.

5. See Table II, page 45.

(* rounded to nearest $.25/MBF)

Chapter 7

2. a. 150 MBF
   b. St & Btr, or 2 & Btr, 2x4 random length
   c. SPF, Hem/Fir, or ES/LP.
   d. paper wrapped, ES/LP, Hem/Fir, 150 MBF, SPF, 2 & Btr, gradestamped, R/L 8"-20", loaded flatcar, S4S, eased edge, DET, Stand & Btr, S-Dry,

Appendix

1. 24+50+48+126+96+36+20 = 400

   (2 x 4 x 400 x 294 x 1)/12 = 78,400 MBF

2. $167.00/MBF*

3. First of all, you realize that the customer is requesting more lumber than will fit on a van or truck, so you can either scale the quantities back proportionately, or in one item, which ever the customer desires. If the lumber is going to weigh 2050 lbs/MBF, and the truck or van can haul 48,000 lbs, a maximum load is going to be about 23 MBF.

Using 23 MBF as the footage, the van freight/MBF is calculated to be $2350/23, or $102.25/MBF. Add to this the mills $5/MBF upcharge for van loading, and the total van freight/MBF becomes $107.25/MBF.

Using the same 23 MBF footage, the truck rate/MBF, which is quoted as $5.83 on 48,000 lbs. minimum, is calculated as:

$5.83 x 480/23 MBF = $121.75/MBF

Having deciphered the freight, you find that a van is cheaper than a truck by $14.50/MBF. Therefore, you will quote this inquiry as a van shipment.

Your quotation price to the customer will be calculated as follows:

1x4	$480.00	(mill price of lumber)
	6.00	(paper wrapping)
	107.25	(van freight)
	+ 20.00	(your profit)
	$613.25	Delivered sell price/MBF

Using the same formula as above for the 1x4, the delivered sell price for the 1x6 is $693.25/MBF, the 1x10 is $673.25/MBF.

(* rounded to nearest $.25/MBF)